Coping with Risk in Agriculture: Applied Decision Analysis

Coping with Risk in Agriculture: Applied Decision Analysis

Editor

Laxmi Dalvi

scitus
academics

Coping with Risk in Agriculture: Applied Decision Analysis

Edited by **Laxmi Dalvi**

ISBN: 978-1-68117-054-1

Library of Congress Control Number: 2015935452

SCITUS Academics LLC,
www.scitusacademics.com
Box No. 4766, 616 Corporate Way,
Suite 2, Valley Cottage,
NY 10989

Printed in United States of America on Acid Free Paper ∞

Contents

Preface

Farmers in developing countries are frequently exposed to the uncertainties of weather, prices and disease. Many farmers live on the edge of extreme uncertainty, sometimes falling just below, and sometimes rising just above the threshold of survival. Farmers do not know whether rainfall will be good or bad over a season; they do not know the prices they will receive for produce sold; and they do not know whether their crops will be infected by disease. These risks are not under the control of farmers but some farmers have developed ways of coping and managing them. Farmers make decisions every day that affect farming operations. Many of the factors that affect the decisions they make cannot be predicted with complete accuracy; this is risk. Farming has become increasingly risky as farmers become more commercial. Farmers need to understand risk and have risk management skills to better anticipate problems and reduce consequences. Decision-making is the principal activity of management. All decisions have outcomes or consequences. However, in most situations the outcome of a decision cannot be predicted. The more complex the risk, the more difficult it becomes for farmers to make an informed decision. For effective decisions to be taken, farmers need information on many aspects of the farming business. Farmers have to find ways of dealing with risk and protecting themselves from the uncertainties of the future.

Editor

Chapter 1

Risk Navigator SRM: An Applied Risk Management Tool

Dana L. K. Hoag and Jay Parsons

Department of Agricultural and Resource Economics, Colorado State University, Campus 1172, Fort Collins, CO 80523-1172, USA

ABSTRACT

Risk Navigator SRM is a ten-step risk management program for agricultural producers, which is based on the strategic planning process. The ten steps are designed to integrate disparate and difficult risk management concepts into a single system that is easy to use, yet still effective. With the aid of computers and the internet, producers can work through each step toward a final comprehensive plan. The website includes 25 decision tools that help producers accomplish each step and provides links to complementary educational programs, like a national agricultural risk education library, the award-winning risk management simulation program called Ag Survivor, and a recently published book that describes the program and provides additional depth and explanations. The ten-step program has been presented in over 200 workshops with over 90 percent approval by participants. The website has averaged over 1,000 unique visitors per month from 120 countries.

RISK NAVIGATOR SRM: AN APPLIED RISK MANAGEMENT TOOL FOR AGRICULTURE

Risk management technology has drastically outpaced the ability of most practitioners to adopt innovations. Studies of farmers and ranchers, for example, consistently show that they do not utilize what is available; some even seem to virtually disregard risk altogether [1, 2]. One survey of nearly 1,000 farmers for Farm Futures [3] found that only 5 percent use available tools. That survey also showed that those that did manage risk tended to be high-end managers with skills not typical of the industry. Agricultural producers were said to resist change because there are too many decisions and too little time.

Congress considered agricultural risks so important that in 1996 it created the Risk Management Agency (RMA) in the US Department of Agriculture. Its purpose is to promote, support, and regulate sound risk management solutions for agricultural producers [4]. According to RMA, there are at least five major forms of agricultural risks [5]: production, market/price, financial, institutional, and human. As shown in Table 1, agricultural producers face many sources of risk and a multitude of ways to manage them across and within these five categories. Price risk, for example, can be affected by product quality, exogenous supply, and government policies; it has at least seven management options, including futures, forward pricing, and storage. RMA efforts have substantially boosted the output of risk information and education to address these risks.

Table 1: Risk management sources and controls; see source in [6]

Risk	Defined	Sources	Management Controls
Production	Uncontrollable events such as weather, pests, or disease make yields unpredictable. Changing technology makes a manager or capital obsolescent. Inputs are unavailable or low quality	Weather, extreme temperatures, pests, disease, technology, genetics, inputs (availability, quality, price), equipment failure, labor, ...	Diversification, insurance (crop, revenue), buildings, storage, vaccines, extra labor, production contracts (e.g., ensure input supply and quality), new technologies (e.g., automate watering)
Marketing and Price	Prices of inputs or outputs change after a producer commits to a plan of action. Price fluctuations stem from domestic and international supplies or substantial changes in demand	Product quality (genetics, disease, handling, input/feed) Product price (quality, timing, global market, weather, government policy, contracts, ...)	Futures and options, forward contracting, retained ownership, quality controls, storage (timing), cooperatives, niche/ value-added marketing, ...
Financial	Stems form the way a business is financed Borrowed funds leverage business equity but increase business risks	Market, production, legal and human risk, interest rate changes, natural disasters (drought), land market changes, foreign exchange, loan calls, ...	Cash reserves, equity, borrowing capacity, reducing other types of risk (production, marketing, etc.), insurance

Institutional	Government or other institutional rules, regulations, and policies effect profitability through costs or returns	Taxes, contract disputes, regulations, government policies, law suits, ambiguous and/or unwritten agreements, neighbors, environmental programs, ...	Estate planning, tax planning, contracts, bonds (e.g., environmental liability), research and education about local laws, ...
Human Resources	The character, health, or behavior of people introduces risk. This could include theft, illness, death in the family, loss of an employee, or a divorce for example	Ambiguous and/or unwritten agreements, poor planning, miscommunication, health or other family disasters, ...	Family planning, including labor planning, clear contracts, training and goal setting, communication, estate planning, ...

While there is little doubt about its importance for decision making, the "challenge is to know how to describe, measure, and communicate risk" [7, page 4]. Consider the parallels to understanding and using probability, which is itself an important component for risk management. Myerson [8] concludes that there is a disconnection between theory and practice because formulas traditionally taught in probability courses are hard to apply to real problems. He suggests that recent advances in computer technology can help overcome these disconnects. Aven [7] goes further by suggesting that a common unifying framework is also needed. For example, websites now combine a framework and computer technology to make managing a stock portfolio relatively simple, even for those people with only a minimal understanding of price analysis. These frameworks require integrating multiple fields, like economics, finance, and statistics, and finding an acceptable balance between the precision and usability.

Agricultural economics and related fields have contributed greatly to developing innovative and effective tools for managing risk in agriculture. But, as is the case with probability, these sometimes disparate theories and concepts can be difficult to understand, which may make integration difficult. Numerous books, articles, and materials are available but are generally inaccessible except by specialists [2]. A 2007 study in Nevada, for example, showed that after six months only 50% of program participants planned to incorporate what they had learned in a risk management workshop [9]. A 2007 study [10] showed that older producers had less knowledge about risk management tools, compared to younger producers, and had less interest in learning more about them. This supports the Farm Futures survey [3] in that many producers view the human capital investment required to learn how to properly incorporate the use of risk management tools into their operation as significant.

The purpose of this manuscript is to describe a new framework for risk management called Risk Navigator SRM. The program is too involved to fully describe here, and it was prepared for education and extension programs, rather than basic research. Nevertheless, researchers might be interested in how disparate parts were integrated to strike a balance between precision and usability, with the purpose of making the components of risk management more usable through a synergistic and reinforcing framework. Precision can interfere with usability and vice versa. That balance in Risk Navigator SRM is based on interaction with producers at over 200 meetings in over a dozen states. This includes a description of supporting software tools made available by website. Our focus here is on showing how probability is integrated with other risk concepts to make risk management more accessible to producers. All ten steps of the process were fully applied to a case study, EWS Farms, which also may be of interest. EWS Farms produce primarily corn and wheat in Northeastern Colorado [6].

We proceed with a description of Risk Navigator SRM and explain the SRM process, which has ten steps. A brief summary is provided for all ten steps, but a more complete description is provided about steps viewed to be of more interest to this readership. We also provide examples of the computer tools available to help producers with each step where appropriate.

RISK NAVIGATOR

Risk Navigator SRM is a program developed to make risk management accessible to typical agricultural producers. The process is general enough that it can be applied to other applications, but all of the examples and tools are customized for agriculture. SRM stands for Strategic Risk Management. Strategic planning [11,12] is an umbrella framework used to organize and integrate risk management concepts and tools for farmers and ranchers. The SRM process has been taught to hundreds of farmers and ranchers under the brand name Risk Navigator SRM or under previous incarnations branded "RightRisk". Risk Navigator SRM is housed on a public website at http://www.RiskNavigatorSRM.com/. The website includes detailed descriptions about a ten-step strategic risk management program and 25 customized computer tools to help decision makers with each step. The tools are matched to the steps and standardized through Xcelsius flash files from Excel where possible. Some files are pdf or Excel spreadsheets. The site also includes links to complementary educational materials, such as a risk practice simulator called Ag Survivor, http://www.AgSurvivor.com/, a comprehensive book titled Applied Risk Management in Agriculture [6], and a general education website called RightRisk, athttp://www.RightRisk.org/. The book includes a comprehensive discussion about risk management, matched to the ten steps, and includes a case study to demonstrate each step. The RightRisk.org website has publications, workshops, and access to resources such as the National Ag Risk Education Library.

THE STRATEGIC RISK MANAGEMENT PROCESS

The framework chosen to make risk management more accessible is strategic planning [6]. The foundation of strategic planning is based on three major phases: strategic, tactical, and operational [12, 13]. The strategic phase of strategic planning is designed to set boundaries based on resources, opportunities, and threats, and to set goals. The tactical phase is designed to evaluate various alternatives for reaching the strategic vision and goals and to develop a solid plan to achieve them. Implementation, evaluation, and replanning occur in the operational stage.

The strategic, tactical, and operational phases have also been linked to agriculture in previous studies. For example, Aull-Hyde and Tadesse [14] modeled strategic differences when long- and short-term risks are considered in agricultural production. Fountas et al. [15] modeled the effects of information flows for strategic, tactical, and operational differences in the context of precision agriculture. Other models review specific details of tactical or operational implementation, while assuming that the strategic phase of the model has already been conducted (Ahumada and Villalobos [16]).

Not surprisingly, risk management researchers have proposed frameworks that intuitively capture the strategic planning process. For example, Clemens and Reilly [17] propose the following steps that are typical of many risk researchers (e.g., [2, 18, 19]):

- identify decision,
- identify alternatives,
- decompose the problem,
- choose best alternative,
- conduct sensitivity analysis,
- repeat if necessary.

These are the same principles found in the tactical and operational phases of strategic planning. The strategic phase is more commonly addressed separately through risk preference and tolerance (e.g., [17, 19]).

Risk Navigator SRM provides formality to agricultural risk management and condenses this information in a way that is understandable to agricultural operators by fitting risk management into the strategic management framework. Hoag [20] developed ten steps that map existing risk management concepts into the three phases (Figure 1). There are three steps in the first phase: (1) determine financial health, (2) determine risk preference, and (3) establish risk goals. These steps are not typically covered in risk management frameworks, but were inspired by the strategic planning process. The first two of these steps were chosen to set boundaries on tactical choices, which are then used to set goals in Step 3, as set forth by the strategic planning process.

Figure 1: Strategic risk management process; see source in the study of Hoag in [6].

There are four steps in the next phase, the tactical stage, which constitute Steps 4–7 in the SRM process: (4) determine risk sources, (5) identify management alternatives, (6) estimate risk probabilities, and (7) rank management alternatives. This tactical phase is based on another framework, the payoff matrix (Table 2), which is commonly used in risk analysis to capture frameworks like that shown above [17, 18]. Each of these steps is designed to elicit a component of a payoff matrix. The payoff matrix is a construct that displays payoffs, usually profits, by management actions (e.g., cash sale, contract sale, hedging on the futures market, etc.) and states of nature (e.g., normal weather resulting in a normal crop and typical crop prices, or bad weather, resulting in a short US crop and high crop prices). Probability is displayed next to each state of nature. The matrix is designed to show risk dimensions in a way that helps decision makers rank risks based on their risk personality, which is further described in Step 7. Summary statistics can be displayed at the bottom of the table to provide more information, such as expected value and standard deviation. For the purposes of illustration, the EWS Farms case study manages corn price risk. There are three marketing management alternatives: selling on the cash market, forward pricing, or hedging. The source of risk is the likelihood of a short U.S. crop.

Table 2: Payoff matrix for EWS farms' corn pricing decision; see source in [6]

			Whole Farm Returns for Management Alternatives		
			no. 1	no. 2	no. 3
Outcome			Forward		
Risk states	$ /bushel corn	Probability	Cash market	Contract	Hedge
Normal crop	$2.10/bu	.35	$287,700	$305,900	$298,200
Short crop	$2.50/bu	.65	$342,500	$332,700	$339,000
Expected value			$323,320	$323,320	$324,720
Standard deviation			$38,749	$18,900	$28,850

One limitation of the payoff matrix is that it only addresses one risk at a time. It can accommodate complex problems with multiple management alternatives, but only one source. Our attempts to discuss joint distributions in risk training workshops, such as price with yield, reduced comprehension and acceptance by producers when presented; therefore we chose to focus on addressing multiple management options for a single risk. Joint distributions and extensions of the model are described by Hoag in [6]. In addition, Ag Survivor risk simulations are based on joint distributions, where appropriate.

The last phase, the operational stage, utilizes three steps intended to carry out the plans made in the tactical stage: (8) implement plans, (9) monitor and adjust, and (10) replan. The first step focuses on the day-to-day activities to assure that good planning efforts are carried out. Monitoring can provide the information needed to determine whether plans should be adjusted. Re-planning takes the decision maker around the circle to start over.

THE STRATEGIC PHASE

We proceed with a brief description of each step. An application to the case study and examples of tools available at the RiskNavigatorSRM.com website are included, where space is appropriate for this audience.

Step 1 (Determine Financial Health): The first step is to determine financial health in order to determine a person's financial capability to take on risks. Financial health refers to assessing the well-being of a business's financial resources, with respect to their ability to take on risk. Educational programs about financial management are widely available in agricultural extension programs. There are six tools available athttp://www.RiskNavigatorSRM.com/ that were based on an extension program in Montana developed by Duane Griffith (http://www.montana.edu/softwaredownloads/financialmgtdownloads.html). This includes typical tools to develop not only commonly used financial statements, like a balance sheet and cash flow statement, but also tools specifically designed to help people understand how health affects risk resilience. Specifically, the "RDfinancial" tool and "Sweet Sixteen Ratio Analyzer" tool identify strengths and weaknesses of a decision maker's financial position. RDfinancial and the Ratio Analyzer provide a plethora of financial information, including the sixteen financial ratios commonly used to describe financial health. RDfinancial also contains a credit scoring model.

Step 2 (Determine Risk Preference): The second step involves assessing a person's risk preferences, which affects a person's attitude about taking on risks. There are many limitations regarding the elicitation of risk preferences [17, 18] but, sometimes, it is worth living with these problems if producers need the extra information at the decision margin. The Risk Navigator SRM's "Risk Preference Calculator" tool offers three different methods to help people gauge their preferences. One method, shown in Figure 2, computes a relative Pratt-Arrow risk preference score [21, 22]. The coefficient of absolute risk aversion, $r_a(w)$, is the negative of the second derivative of utility, U, for wealth, W, divided by the first derivative of utility for wealth:

$$r_a(w) = -\frac{U''(w)}{U'(w)} \quad (1)$$

This coefficient is positive for risk-averse individuals, zero for risk neutral individuals, and negative for risk-loving individuals. People that are risk averse would pay a premium to avoid risk. Risk-neutral individuals maximize expected values and ignore risk. Since "r" changes with the size of the gamble, the concept of a coefficient of relative risk aversion (RAC) was created simply by multiplying the coefficient of absolute risk aversion by wealth:

$$RAC = Wr_a(w) \quad (2)$$

The RAC equals 0 for someone that is risk neutral. It varies from about 0.5 to 4.0 for risk-averse people, as suggested by Hardaker et al. [18]:

- hardly risk averse at all,
- somewhat risk averse (normal),
- rather risk averse,(iv): very risk averse,
- almost paranoid about risk.

A common, yet limited, functional form used for utility is the negative exponential utility function:

$$U(w) = 1 - \exp^{(-rW)} \quad (3)$$

This is a convenient functional form for illustration since the coefficient of absolute risk aversion is r. The RAC can be found by eliciting a person's utility function, which can be estimated by the ELCE (Equally Likely Certainty Equivalents) method as described by Hardaker et al. in [18] (Figure 2). ELCE elicits equally likely certainty equivalents by asking a series of questions that present 50-50 bets, which can be used to sketch a utility function like that shown in the lower left quadrant of Figure 2. The certainty equivalent is the certain amount that a person would be indifferent to receiving compared to an expected value with risk. To provide a more realistic scenario, onion production was chosen for the ELCE questions in the Risk Preference

Calculator tool, since the crop tends to produce boom or bust returns (so we could use the 50-50 technique). These "bets" are presented by slider bars as shown in Figure 2. This allowed the program to remain simple and realistic, and fit well with a 50-50 bet for agricultural producers. Using the Hardaker et al. scale, the EWS case study farmer turned out to be quite risk tolerant, with a risk preference score of 0.52 (somewhat risk averse). This score is based on derivatives from the utility curve fitted in the lower left quadrant of Figure 2.

The second risk preference assessment method (not shown) in the Risk Preference Calculator provides users the option of taking a short risk quiz designed by Grable and Lytton [23]. The quiz asks 13 questions, which are tabulated to provide a score from "low risk tolerance" to "high risk tolerance." More than one method is offered to counteract inherent difficulties in measuring risk preferences. In this case, psychological research offers a completely unique approach in the form of this quiz.

The third method links risk preference to risk tolerance, which is the amount $X, where a person would be indifferent between an equal chance of receiving $X and losing $\$\frac{X}{2}$ [8]. A risk-averse person that would pay a certainty equivalent of $7,000 for a risky, 50-50 bet of receiving either 0 or 20,000, for example, has a risk tolerance of $15,641; risk tolerance increases to $99,833 for someone willing to pay $9,900 for that same bet. A risk-neutral producer would of course be willing to pay $10,000. Risk tolerance uses a different means than ELCE to elicit tolerance, but is closely linked to preference. Relative risk tolerance (RRT) can be derived from the RAC, since absolute risk tolerance, R, is the inverse of absolute risk preference, r in the negative exponential utility function. The scale of RRT spans from 0.25 (1/4) for a person who is almost paranoid about risk to 2 (1/0.5) for someone who is rarely risk averse to 10 (1/.1) for someone that is almost risk neutral. This allows the Risk Preference Calculator to provide a comparable estimate of the relative risk preference based on risk tolerance, which opens the door to an entirely separate and more prevalent literature. For example, Howard [24] defined R for firms that he looked at in terms of annual sales, equity, and income. He found that R=1.24 multiplied by net income, or 6.4% of sales, or 15.7% of equity for the businesses that he examined. The Risk Preference Calculator can therefore provide a parallel estimate of risk preference by simply asking the producer for net income, sales, or equity.

Utilizing three methods to elicit risk preferences makes the Risk Preference Calculator tool more accessible to users with differing tastes. It also provides some continuity and comparability for those users willing to apply all three methods, which helps combat the inherent problems with estimating risk preferences [17].

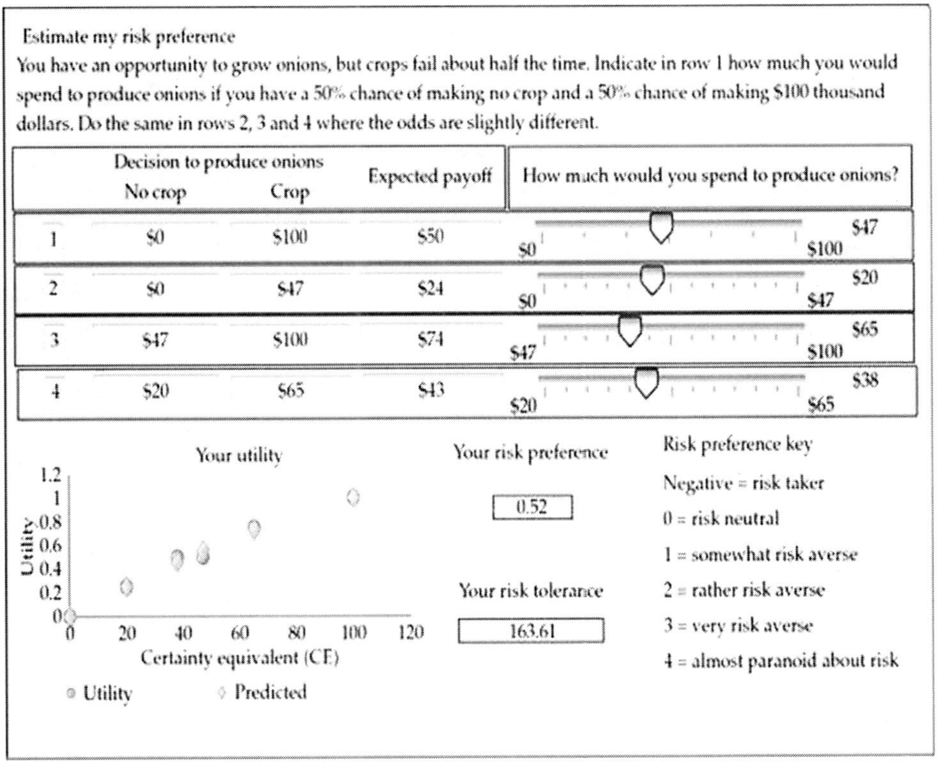

Figure 2: Elicitation of risk preference/tolerance in Risk Preference Calculator. See source in [6].

Step 3 (Establish Risk Goals): There are seven goal planning tools on the Risk Navigator SRM website, including action planning, team roster, mission statement, time management, and transition planning. These classic tools are updated to assist decision makers with developing their goals in the "Strategic Goal Worksheet." EWS Farms' strategic goals were [6] the following.

- Strategic Goal no. 1 (Financial): Ensure short- and long-term financial success by maintaining business profitability, while expanding the overall business financial resource base.
- Strategic Goal no. 2 (Family): Continue to live, work, and grow our families in a rural, agricultural environment. Encourage individual development and exploration in a manner that is consistent and flexible in order to allow all individuals to reach their full potential.
- Strategic Goal no. 3 (Organizational): Continue to pursue organizational structures that fit the family dynamics of the operation, as well as allow for strategic goal attainment. Also, increase the business activities efficiency of the operation.
- Strategic Goal no. 4 (Integrated Farm Management): Manage our farm as a cointegrated unit, while providing a step-by-step process for developing a strategic risk management plan.
- A comprehensive description of all of their goals and mission statement can be found in Applied Risk Management in Agriculture [6].

THE TACTICAL PHASE

Step 4 (Determine Risk Sources): The four steps in the tactical phase are those required to build a payoff matrix. The first step is to determine risk sources. In addition to determining risk sources, it is also important to prioritize risks so that management efforts can be focused. EWS Farms chose to focus on price risk for corn. This was interesting since it was not even one of their risk management goals. However, it is not inconsistent with observations at risk education extension meetings. Men that have attended the Risk Navigator SRM workshops have overwhelmingly chosen price and yield risk (revenue) over all other types of risk when doing this exercise. Decision makers are encouraged in this step to review their goals and information with those available in the study by Hoag in [6], like those shown in Table 1.

Several methods are demonstrated in the book titled Applied Risk Management in Agriculture [6], such as influence diagrams and SWOT ((S)trengths, (W)eaknesses, (O)pportunities, and (T)hreats) analysis. However, the method chosen for the Risk Navigator tool, "Risk

Influence Calculator," uses a risk-influence diagram to help decision makers sort out and prioritize risk. The Risk Influence Calculator is simple and only asks producers to list each risk, then to rate it on a scale of 1 (low) to 10 (high) for (A) probability of occurring, (B) impact if it occurs, and (C) influence to stop it from occurring.

After reviewing his goals and reading what others said about risks (e.g., Table 1), EWS Farms identified the following strategic risks:

- Market/Price
  1. Corn Price. Will my price cover my costs?
- Production
  1. Weather. Will rainfall support my crop stand?(
  2. Hail. Will hail destroy half my crop?
  3. Input (seed). Will good corn seed be available at a reasonable price?
- Financial
  1. Expansion. Can the operation generate enough profit to cover new land payments?
- Human
  2. Family. Will my dad retire?
- Institutional
  1. Water. Will irrigation water be restricted?

All risks identified are placed in a graph that plots influence against risk, so the decision maker can prioritize risks where the biggest impact can be made. The Risk Influence tool has been one of the most popular in Navigator risk management workshops, as it creates a lot of discussion and rethinking about priorities.

Step 5 (Identify Management Alternatives: The book, Applied Risk Management in Agriculture [6], describes four main techniques to manage risks: (1) avoid it, (2) transfer it, (3) assume or retain it, or (4) reduce it. To keep things understandable, we use a risk profile. A risk profile is a multidimensional representation of the consequences that stem from each management action. For simplicity, we use a probability density function (PDF) as the risk profile, since it is familiar to most people and it contains information that is relevant to managing risk, such as mean, mode, maximum, and minimum. We can show the consequences of the four basic management actions mentioned above through simple manipulations of the PDF such as skewing, truncating, changing variance (squishing), or changing the mean (moving). This simple representation of a risk profile (and terms like squishing) is meant to build basic skills and understanding in the participants; however, perhaps more importantly, the graphical depictions engage the audience in what may often be perceived as a dry subject. It also ties the concept of management alternatives to PDFs for the next step concerning likelihood. Like the previous step, the book also provides an extensive list and brief discussion about specific techniques commonly used in agriculture, like crop insurance and the futures market (e.g., Table1).

Decision makers at EWS Farms identified the following possibilities to manage price risk in their corn:

- Cash market sales (selling everything in the market at harvest),
- Forward contracting to a local grain elevator (contracting for a fixed price before harvest),
- Hedging on the futures market (selling a contract on the futures market for a fixed price),
- Spreading out crop sales across the year (multiple marketing points),
- Maintaining flexibility on timing of sales (storing, then selling opportunistically).

After more consideration, EWS Farms chose three management alternatives: cash market, forward contracting, and hedging. They developed a comprehensive marketing plan using the "Marketing Plan" tool in Risk Navigator SRM. Six components exist in this marketing plan, and each component has its own worksheet: (1) The Relationship between the Strategic Risk Management Plan and the Marketing Plan; (2) Production History and Expectations; (3) Expected Prices; (4) Production Costs; (5) Price, Date, and Quantity Targets; (6) Review and Evaluation. Each of their marketing alternatives is carefully planned, including the distribution of the sale over time. The marketing alternatives are also based on ten years of local data on EWS Farms and for their local elevator (including basis adjustments for prices).

To help EWS Farms prepare for the remaining steps, the book recommends putting their information into a decision tree and then a payoff matrix. The main value of starting with the decision tree, shown in Figure 3, is in its visual construction [25], which requires the decision maker to identify all relevant courses of action, events, and payoffs in a clear and understandable manner. It also makes it easier to process information to put into a payoff matrix, as shown in Table 2.

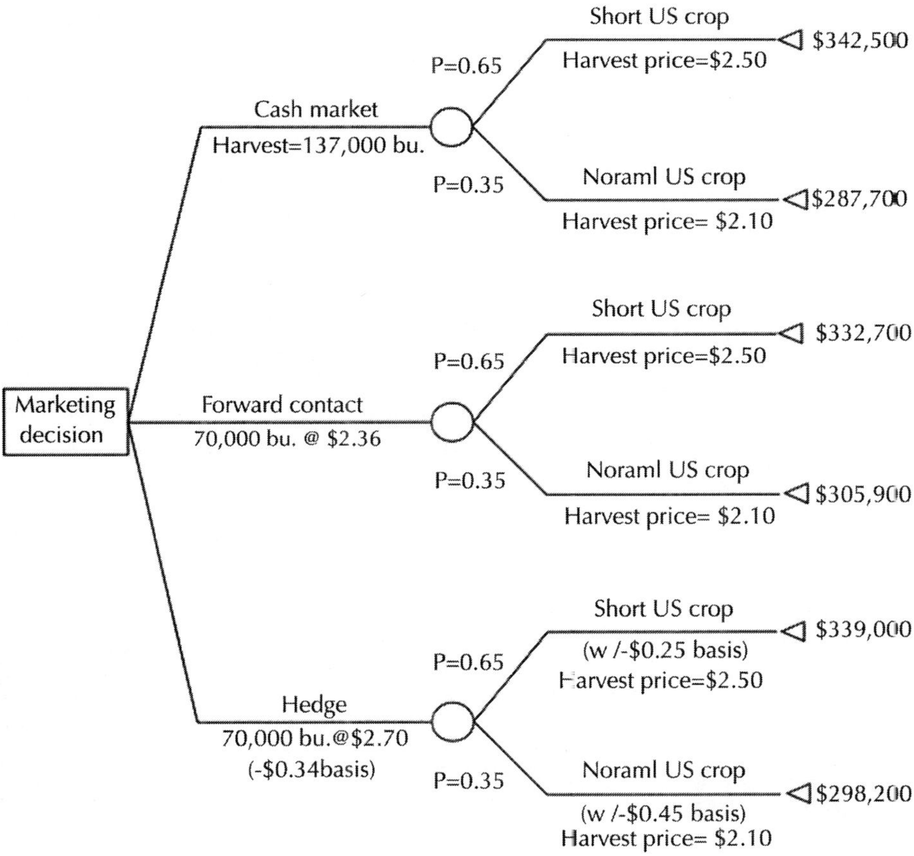

Figure 3: Decision tree for EWS Farms corn pricing decision; see source in [6].

Step 6 (Estimate Probabilities): The concept of probability has been cultivated throughout the ten-step process by encouraging decision makers to represent risk with a "risk profile" and to think of a risk profile in terms that can be easily understood from the basic shape of a PDF. They are encouraged in Step 2 to determine their "risk personality" (preference) so they can find the risk profile that best suits them. In Step 5 they are shown how a risk profile is affected by risk management alternatives, where the random variable is usually income or cost. In this section we show how to tie the concept of the risk profile to a probability density function (PDF) or a cumulative density function (CDF) more directly. Much of the book chapter [6] is dedicated toward a basic lesson about the PDF and CDF, including concepts like mean, mode, median, variance, standard deviation, and coefficient of variation. Also included are descriptions about how to interpret basic shapes of the PDF. The discussion is very basic, aimed at making the PDF and CDF concepts that people could use to quickly and intuitively interpret the basic statistical components that are important for risk management (e.g., measures of central tendency, spread, and range). For example, it is easy to convey that management tools like insurance or the futures market "squish," "move," or "truncate," a PDF.

The "Risk Profiler" tool makes it relatively easy for producers to build probability density functions (PDF's), which then can be used to provide information for probability in the payoff matrix. The "art" of eliciting probability is fraught with limitations (e.g., [17–19]), but the PDF's are used here to simplify the process of estimating probability in the payoff matrix and to provide a stable mechanism to tie the steps together. Furthermore, Hoag [6] discusses some of the common problems, like anchoring, and how to avoid them.

There are at least two major ways to elicit probability information (6,10): asking the expert about his or her degree of belief about an outcome or present the expert with a lottery that reveals their probability values. Risk Profiler provides three options to elicit a PDF by asking the expert about what they believe will happen. The most straightforward method assumes that the future will look like the past ten periods. Figure 4 shows ten annual corn price entries for EWS Farms and resulting PDF, CDF, histogram, and summary statistics. The PDF is assumed to be Normal for simplicity. However, for increased education, a histogram is also drawn to help decision makers understand what might be hidden behind the normalized function. For example, a bimodal distribution is hidden by the normal distribution in the case of corn price on EWS farms. A few summary statistics are also provided for each estimation method.

The second method offered by Risk Profiler to elicit a PDF involves having the user "describe profile features." In this case, a PDF can be drawn based only on the minimum, most likely, and maximum values elicited from the decision maker for the random variable. The PERT distribution is applied as follows:

$$\mathrm{PERT}(a,b,c) = \mathrm{Beta}(\alpha_1 \alpha_2) * (c-a) + a \qquad (4$$

where "a" is the minimum value, "b" is the most likely value, "c" is the maximum value, and "Beta" is the beta distribution [19].

Furthermore,

$$\alpha_1 = \frac{(\mu - a) * (2b - a - c)}{(b - \mu) * (c - a)} \qquad (5)$$

$$\alpha_2 = \frac{\alpha_1 * (c - \mu)}{(\mu - a)} \qquad (6)$$

In this case, the resulting PDF and related information are pictured on a screen exactly like that shown in Figure 4, but the upper left quadrant is replaced with a section to collect the PERT input data.

The last method is "describe profile PDF." In this method, people provide five sets of probabilities and values for the random variable, see Table 3.

This is a variation on the fractal method [17, 18]; in eliciting the PDF for prices for a given crop, for examp e, a decision maker might be asked to pick several price values and give an associated probability for each one. Asking for the producer to supply the probability and values simultaneously is a combination of what Frey [26] called the fixed value method and fixed probability method. The fixed value method asks an expert the probability that the actual value is higher (or lower) than some arbitrary number for the random variable. The fixed probability method has the decision maker specify the value range for a random variable that corresponds to a specific probability. The Risk Profiler method lets the user simultaneously mix and match the fixed value and fixed probability methods by allowing them to enter either probabilities or values.

Table 3

Probability	Price
.30	$2.00
.20	$2.25
.40	$2.50
.20	$2.75
.10	$3.00

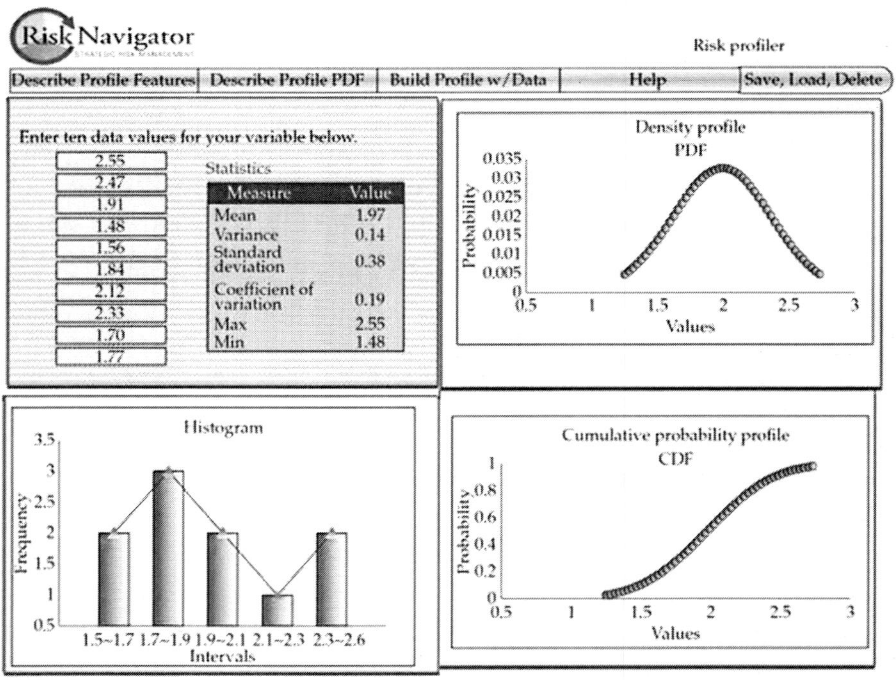

Figure 4: Building a PDF with Risk Profiler; see source in [6].

Step 7 (Rank Management Alternatives): The final step of the tactical stage is to rank the various alternatives considered to this point and select those with the most desirable outcomes. Risk Navigator

SRM offers three tools to help decision makers rank risks. "Value at Risk" (VaR) is a popular method for capturing the downside risk in financial decision making. It is an evaluation of what you stand to lose with an investment. VaR answers the questions "What is my worst-case scenario?" or "How much could I potentially lose in a really bad month?" [27]. This strategic tool considers only the undesirable parts of dispersion, those on the negative side of the mean, as opposed to the standard deviation, for example. The VaR tool is simple to use and involves only one screen (not pictured here). A second tool, the "Risk Efficiency Tool," uses Stochastic Efficiency with Respect to a Function to rank outcomes for all levels of risk preference, except risk preferring.

The easiest and most effective tool, "Risk Ranker," uses the payoff matrix to link risk personalities with the risk profiles. This tool allows a user to directly compare risk profiles, which can provide a lot of information by itself, and offers an instant comparison of the management alternatives under consideration with seven ranking rules that cater to different risk personalities (e.g., someone that is avoiding risk as compared to someone that wishes to maximize expected value).

After filling out a payoff matrix for up to five management alternatives, the program can be used to compare risk profiles. For example, on the second tab, "Compare Profiles," shown in Figure 5, the payoff matrix entered by the decision maker is reprinted in the upper left corner. The first column paired with any management alternative (e.g., cash) replicates the information entered in "Risk Profiler," which provides continuity to the program and reinforces how the risk profile integrates with the payoff matrix. All five PDFs and CDFs for EWS Farms are plotted in one graph and summary statistics are provided in tabular form. Decision makers are provided with information about how to rank alternatives with methods that use only distributions, such as stochastic dominance [6], and may therefore use this method alone to rank risks.

In many, if not most, cases risks cannot be ranked by visual inspection of the PDF or CDF. Therefore, the next tab over, "Risk Ranker" displays seven different risk ranking measures for the payoff matrix, all on one screen: Maximize EV, Maximax, Most likely, Minimax regret, Hurwicz, Maximin, and the Laplace Insufficient Reason Index (Table 4). Each of these techniques ranks risks based on different aspects of the payoff that might, or might not match a decision maker's risk personality. For example, Maximize EV (expected value) is for risk neutral producers. Maximin chooses the alternative with the maximum, minimum outcome and is therefore designed for very risk-averse individuals.

Table 4 shows that each of the rules focuses on different areas. None is comprehensive, and many are oversimplified. For example, Maximin would choose A in the two five-year income streams A and B shown below, since it focuses only on the minimum; however, most people, if not all, would choose B if they could:

$$
\begin{aligned}
A &= (10,10,10,10,10) \\
B &= (100,100,100,9,100)
\end{aligned}
\tag{7}
$$

The table presented in the tool makes it handy to compare rankings quickly and easily so that many dimensions can be considered, and is very effective at getting across the message that people need to match their risk management personality to their risk ranking techniques.

Table 4: Attributes of basic decision rules[a]

Decision Rule	Mean	Variation	Low	High	Other
Maximized Expected Value—Choose alternative with highest expected value	x				
Maximax—Choose alternative with best outcome				x	
Most Likely—Choose alternative that is most likely to occur (subjective)					x
Minimax Regret—Choose alternative with least hindsight regret		x	x		
Hurwicz—Weighted average of maximax and minimax			x	x	
Maximin—Choose alternative with best of the worst outcomes			x		
Laplace—Simple average	x				

[a]See a more detailed description in [6]. An x is placed in the column where the rule is primarily focused.

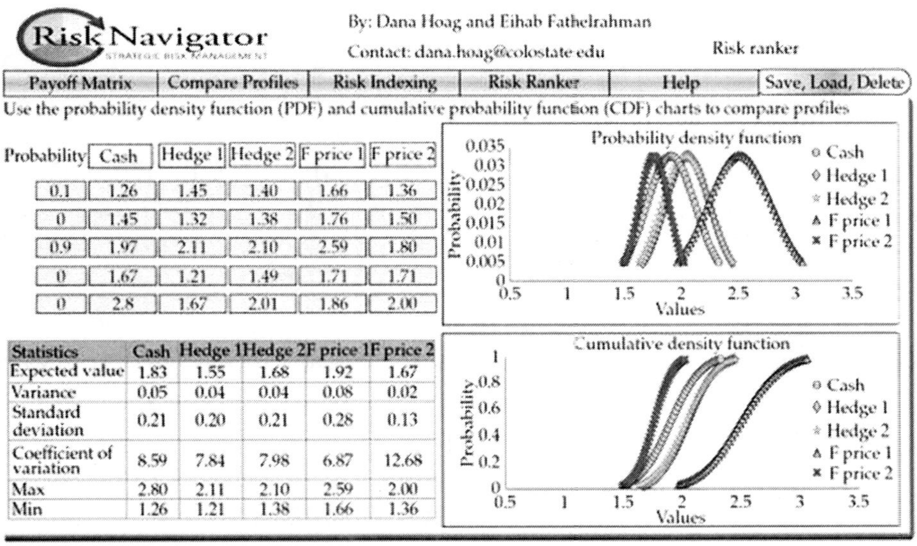

Figure 5: Risk Ranker—compare profiles—see source in [6].

THE OPERATIONAL PHASE

Steps 8.10

The operational stage involves putting plans into action. The first step, implementation, involves acquiring the necessary resources, scheduling the tasks to be completed, and overseeing all aspects of the plan [6]. As plans are implemented, the second step in this phase kicks in as resources need to be monitored so adjustments can be made as needed. The last step in the phase, and step ten overall, is re-planning. If not actively brought to the attention of those involved with the program, this step may be easily ignored by many managers since it could highlight what was not achieved. Re-planning is also the first step in preparation for starting the cycle over. There are five SRM tools available on the website for the operational phase: (1) Action Planning Worksheet, (2) Critical Success Indicator Worksheet, (3) Resource Flow

Plan, (4) Risk Management Worksheet, and (5) Time Management Worksheet. A detailed description of how EWS Farms applied this step is provided by Hoag in [6].

CONCLUSIONS

Risk Navigator SRM integrates risk management techniques into one place and provides farmers and ranchers the resources to use the products from their home office. It is both a learning tool as well as a means to help producers actually manage risks, financial and other, in their operations. It is also very practical for students in economics and business. The manageable steps allow for farmers and ranchers to learn the tools at their own speed, while providing the opportunities to customize the data for their own farm/ranch. We find that the difference between this and other programs designed to manage risk is the integration of basic risk management principles into a structured and easy to learn format. This allows people to use concepts that have been individually available for decades, but inaccessible, because they are most valuable in a framework, which typically requires expertise to build. The book, Applied Risk Management in Agriculture [6], can supplement the website by providing detailed descriptions of each step and by providing additional educational opportunities. Navigator workshops have been presented in formats from 45 minutes to two days. The program has grown and the tools can be continuously upgraded since they are offered at the Navigator website. A blog has been added and a free, ten-step online education program is nearly complete.

ACKNOWLEDGMENTS

The authors thank to the USDA Risk Management Agency for partial funding and the anonymous reviewers for their helpful comments.

1. REFERENCES
2. J. Harwood, R. Heifner, K. Coble, J. Perry, and A. Somwaru, "Managing risk in farming: concepts, research and analysis," Report 774, U.S. Department of Agriculture-Economic Research Service, Market and Trade Economics Division and Resource Economics Division, Washington, DC, USA, 1999.

3. B. Fleisher, Agricultural Risk Management, Lynne Rienner Publishers, Boulder, Colo, USA, 1990.
4. Risk Management Agency—U.S. Department of Agriculture, Building a Risk Management Plan: Risk-Reducing Ideas that Work, U.S. Department of Agriculture, Washington, DC, USA, 1998.
5. Risk Management Agency—U.S. Department of Agriculture, A Risk Management Agency Fact Sheet About the Risk Management Agency, Program Aid 1667-02, U.S. Department of Agriculture, Washington, DC, USA, 2009.
6. Risk Management Agency—U.S. Department of Agriculture, Introduction to Risk Management: Understanding Agricultural Risks: Production, Marketing, Financial, Legal and Human Resources, U.S. Department of Agriculture, Washington, DC, USA, 1997.
7. D. L. Hoag, Applied Risk Management in Agriculture, CRC Press-Taylor Francis, Boca Rotan, Fla, USA, 2009.
8. T. Aven, Foundations of Risk Analysis: A Knowledge and Decision-Oriented, John Wiley & Sons, Chichester, UK, 2003.
9. R. Myerson, Probability Models for Economic Decisions, Thomson, Brooks/Cole, Belmont, Calif, USA, 2005.
10. K. Curtis, "Risk management programming: is it effective at creating change," in Proceedings of the National Extension Risk Management Education Conference, Phoenix, Ariz, USA, April 2007,http://www.agrisk.umn.edu/conference/agenda.aspx?ConfID=4&Output=Sum.
11. G. F. Patrick, A. J. Peiter, T. O. Knight, K. H. Coble, and A. E. Baquet, "Hog producers' risk management attitudes and desire for additional risk management education," Journal of Agricultural and Applied Economics, vol. 39, no. 3, pp. 671–687, 2007.
12. G. Steiner, A Step by Step Guide to Strategic Planning: What Every Manager Must Know, The Free Press, New York, NY, USA, 1979.
13. G. Toft, "Synoptic (one best way) approaches of strategic management," in Handbook of Strategic Management, J. Rabin, G. Miller, and W. B. Hildreth, Eds., pp. 545–560, Marcel Dekker, New York, NY, USA, 2nd edition, 2000.

14. B. Bilgen and I. Ozkarahan, "Strategic tactical and operational production-distribution models: a review," International Journal of Technology Management, vol. 28, no. 2, pp. 151–171, 2004.
15. R. Aull-Hyde and S. Tadesse, "A strategic agricultural production model with risk and return considerations," Agricultural and Resource Economics Review, vol. 23, no. 1, pp. 37–46, 1994.
16. S. Fountas, D. Wulfsohn, B. S. Blackmore, H. L. Jacobsen, and S. M. Pedersen, "A model of decision-making and information flows for information-intensive agriculture," Agricultural Systems, vol. 87, no. 2, pp. 192–210, 2006.
17. O. Ahumada and J. R. Villalobos, "Planning the Production and Distribution of Fresh Produce," Annals of Operations Research. In Press.
18. R. T. Clemens and T. Reilly, Making Hard Decisions with Decision Tools®, Duxbury, Pacific Grove, Calif, USA, 2001.
19. J. B. Hardaker, R. M. Huirne, J. R. Anderson, and G. Lien, Coping with Risk in Agriculture, CABI Publishing, Wallingford, UK, 3rd edition, 2004.
20. D. Vose, Risk Analysis: A Quantitative Guide, John Wiley & Sons, New York, NY, USA, 2nd edition, 2000.
21. D. L. Hoag, "The strategic risk management process," Unpublished proposal, Department of Agricultural and Resource Economics, Colorado State University, Fort Collins, Colo, USA, 2005.
22. K. J. Arrow, Aspects of the Theory of Risk-Bearing, Academic Bookstore, Helsinki, Finland, 1965.
23. J. W. Pratt, "Risk aversion in the small and the large," Econometrica, vol. 32, pp. 122–136, 1964.
24. J. E. Grable and R. H. Lytton, "Financial risk tolerance revisited: the development of a risk assessment instrument," Financial Services Review, vol. 8, no. 3, pp. 163–181, 1999.
25. R. A. Howard, "Decision analysis: practice and promise," Management Science, vol. 34, no. 6, pp. 679–695, 1988.
26. A. N. Rae, Agricultural Management Economics: Activity Analysis and Decision Making, CABI Publishing, Wallingford, UK, 1994.
27. C. Frey, "Briefing paper part 1: introduction to uncertainty analysis," Department of Civil Engineering, North Carolina State University, 1998,http://legacy.ncsu.edu/classes/ce456001/www/Background1.html.

28. D. Harper, "Introduction to value at risk (VaR)," 2008,http://www.investopedia.com/articles/04/092904.asp.

Chapter 2

The Economics and Institutional Economics of Wildlife on Private Land in Africa

Brian A Child[1], Jessica Musengezi[1], Gregory D Parent[1], and Graham F T Child[12]

[1]Stellenbosch Institute for Advanced Studies (STIAS), Wallenberg Research Centre at Stellenbosch University, Marais Street, Stellenbosch, 7600, South Africa

[2]Department of Geography, Center for Africa Studies, University of Florida, 427 Grinter Hall, P.O. Box 115560, Florida, USA

ABSTRACT

In southern Africa, there are now 10,000 to 14,000 private ranchers that promote wildlife enterprises alone or in some in combination with domestic livestock. An important conservation success, this new bio-experience economy also creates social well-being through economic growth and job creation. It is an economic sector that needs to be taken seriously, not least because it pioneers policies that inform the valorization and sustainable management of ecosystem services. The article describes the historical emergence of a sustainable use approach to wildlife conservation since the Arusha Conference in 1963. It suggests that indigenous multi-species systems may have ecological advantages over modern livestock production systems, but these are difficult to quantify in complex dry land ecosystems and are trumped by economics and political processes. However, wildlife provides the foundation for a bio-experience economy that has a decided comparative economic advantage over agro-extractive commodity production (like beef) in dry lands. We describe how new policy approaches, especially the valorization of wildlife and the devolution of proprietorship to landholders and communities, have allowed wildlife's economic advantages to be reflected in land use decisions through both 'game ranching' and 'community-based natural resource management'. Institutional changes have modified the economics of wildlife in dry lands, promoting both conservation and development by allocating environmental raw materials to higher-order goods and services. A further goal of the paper is to describe practical economic methods for assessing and explaining the wildlife sector to policy makers in terms of its profitability, both to individual landholders and to society through jobs and economic growth. The paper covers a 50-year period between the PhD studies of the four authors and takes a trans disciplinary approach which values the knowledge of practitioners as much as the academic literature.

REVIEW

Background

This provides a review of the economic evolution of the private wildlife sector in southern Africa since the 1960s. It is intended to be accessible to a wide readership (not just economists or wildlife managers). It draws on the academic literature but also on considerable practitioner knowledge networks including gray and oral literature. It describes the history of the wildlife sector from the perspective of the economic principles and lessons that are emerging from this policy experiment. Our purpose is to describe the evolution of a sustainable use (or neo-liberal) approach to wildlife management (SASUSG 1996; Martin 2009), which is underrepresented in the literature relative to alternative paradigms of conservation and development.

We evaluate the economics of wildlife using the private sector experience for methodological reasons, although this economy is smaller than the wildlife economy based on Africa's protected areas, less scalable to Africa's extensive dry lands than community-based wildlife management, and under political scrutiny because of its racial history. However, stronger property rights and a gradual shift from administrative to free-market competitive pricing mean that pricing is less distorted and economic evaluation is more reliable on private land. Evaluating the economics of dry lands is difficult at the best of times because internalizing the full costs and benefits of resources like water, grazing, and wildlife is complicated because they are fugitive in space and time, while ecological processes such as carbon and water cycles and soil processes are complex and nonlinear, making the attribution of costs and benefits difficult.

The political economy of natural resources further muddies the economic waters. Historically, wildlife has been monopolized (e.g., protected areas) and heavily regulated by the state, and market signals violate many of the assumptions needed for a sound economic analysis given the recent and partial shift from centralized planning to market-based approaches. Wildlife also coexists with pastoralists and subsistence farmers who are politically marginalized, with limited and contradictory rights (e.g., of exclusion and use) to natural resources. Thus, economic signals are distorted by a combination of state monopolies, regulation, administrative pricing, weak institutions and/or predatory governance, and confused or open-access property regimes, making economic analysis near impossible. 'Market failure', an economic term that refers to the gap between the actual price of a good and its real value, is pervasive.

On private land in southern Africa and, indeed, where private land exists elsewhere in Africa's dry lands, the shift from livestock to wildlife is widely acknowledged but not reliably quantified. There are reported to be some 9,000 'game ranches' in South Africa alone plus over 15,000 which combine wildlife with livestock (Cousins et al. 2008). The rapid expansion of the wildlife sector mirrors the growth in the regional and global bio-tourism industry and is reflective of shifting global terms of trade. The South African game ranching sector has expanded at between 5% and 20% annually in the last decade, whereas real farm incomes have declined by 5.3% (Dry 2010a, b; Jenkins 2011). Game ranching is a diverse sector that combines ecotourism, the sale of live animals, several forms of hunting, and as a by-product, meat production. Prime wildlife areas generate large amounts of income and employment from tourism, while trophy hunting is geographically more extensive and a reliable form of income where wildlife populations are recovering or in areas of limited tourism potential. Hunting is critical in the transition from cattle ranching back to multi-species production systems, except where large amounts of capital are available for restocking, because it utilizes only 2% to 3% of wildlife populations that can biologically expand at 10% to 30%. The markets for hunting also appear to be larger and more robust than is often assumed, and both the supply and price of trophy animals has been rising steadily in southern Africa since the mid-1970s (Booth 2002, 2009; Lindsey et al. in press). For example, elephants sold for US dollars $250 in the late 1970s and are now worth $12,000 to $20,000 in trophy fees and up to $65,000 when we include outfitter fees. This suggests that up scaling of trophy hunting and ecotourism is possible without tumbling prices. The southern African wildlife sector is a tiny share of the global tourism and hunting market, and tourism is a cluster industry that benefits from economies of scale and from trends in global terms of trade. By contrast, beef commodity prices have been stagnating globally for nearly four decades (albeit with a significant upturn in the last two years). Beef production has also shifted steadily away from the dry lands since the 1960s following technological 'advances' like grain feeding, nitrogen supplementation, and feedlots, while Africa's dry lands are unlikely to be competitive with large-scale meat production in Argentina, Brazil, or the USA.

Land use trends on private land, together with the data summarized in this paper, confirm that the bio-experience economy based on the multiple uses of wildlife is an increasingly serious industry. It is beginning to be included in national agricultural statistics, especially in South Africa, though it is not yet well quantified because of its diversity compared to commodity markets. As an example of the magnitude of this sector, private land hunting alone is worth over US $one billion in South Africa. Even the casual observer can see that Africa's famous protected areas have significant impacts on the growth of nearby urban centers like Arusha (Tanzania), Kasane, Maun (Botswana) as well as national capitals.

The purpose of this paper is to present the case for wildlife as a profitable form of land use in semiarid savannas, creating more jobs and economic growth than meat commodity production. For ecological and economic reasons, the bio-experience economy is a legitimate development option. With a more favorable policy environment, it could be applied on a much broader scale than is currently the case, especially if it can be adapted to Africa's communal area circumstances through approaches like community-based natural resource management. Furthermore, these economic advantages are increasing. Global terms of trade are moving in favour of the bio-experience economy; the wildlife sector is a new industry with considerable potential for adding value through product diversification and development, and it is synergistic with ecosystem services like water and carbon that are as undervalued as wildlife was several decades ago.

In focusing this paper on presenting the economic case for wildlife, we are not taking an 'either-or approach' to economic alternatives. Our aim is to maximize the well-being of people living in marginal areas in environmentally sustainable ways, with wildlife conservation contributing to this in areas where it has a comparative advantage. Nonetheless, we cannot avoid the implications that meat commodity production through livestock or wildlife is economically and environmentally questionable in dry lands. White minority regimes supported cattle ranchers with significant subsidies for several decades, but when these were removed in the transition to black majority rule, the beef economy in dry lands was shown to be financially marginal and ecologically hazardous, and many of these landholders only survived by adding or switching to wildlife operations (Child 1988; Grossman and Gandar 1989).

Unlike beef commodity production, traditional livestock systems have multiple values (e.g., milk, meat, cash, and store of value). This may give them financial and ecological advantages, but in the absence of recent or detailed long-term studies of these systems that include the costs of labour and environmental capital, we are safer to assume that they are unlikely to outperform commercial systems by a factor of two. In other words, their viability is still questionable, and we are irresponsible if we do not question the assumption that dryland animal production can meaningfully address poverty. While we claim no expertise in pastoral economics, our livelihood surveys (unpublished data) from Shorobe and Sankuyu communities in northern Botswana suggest that household poverty is higher, and income distribution far more skewed, in communities that depend on livestock (Shorobe) than those with wildlife (Sankuyo). Additional data from communal lands in South Africa (see below) suggest that the economic output communal land is low. While beyond the scope of this paper, rural people in the drylands of southern Africa are vulnerable to extreme poverty and appear to be surviving more on wages, remittances, and government transfer payments than on local production. Generally speaking, livestock are kept primarily for cultural reasons and as a mechanism of savings and coping with risk. It is also significant that they are an asset that can be privately owned and used to harvest natural resources in situations where people generally have limited rights to valorize, manage, or exclude others from 'their' natural resources. This paper is about the wildlife economy and its potential to drive development.

The challenge of scaling up this economic model and adjusting it to the livelihoods of the rural poor in Africa's communal drylands moves beyond the economic model itself to the challenges of transferring it in difficult political, institutional capacity and demographic circumstances. This policy response in southern Africa has been labeled community-based natural resource management (CBNRM). Since the mid-1980s, CBNRM has attempted to replicate the economic success of wildlife on private land in the complex institutional circumstances of Africa's communal areas (Martin 1986; Metcalfe 1993; Jones and Murphree 2004; Child and Barnes 2010) by attempting devolving full rights to use, manage, and benefit from wildlife to communities as had been done for private game ranchers (Murphree 1994), albeit falling short of this intention (Martin 2009). On the one hand, it is naïve to suggest that wildlife can resolve poverty for millions of people in drylands, but on the other, CBNRM has clearly benefited both people and wildlife where it has been implemented with some level of competence (Child et al. 2003; NACSO 2006). However, in several countries where superior wildlife resources could clearly benefit people, this has not occurred (Nelson and Blomley 2009; Nelson 2010). We fall on the side of the debate that the problem lies more with implementation than with the model itself (Hulme and Murphree 2001), which often falls far short of best practice because communities are not really empowered to use, manage, and benefit from wildlife (or for most natural resources that matter) (Martin 2009; Ribot et al. 2010).

Although strongly associated with wildlife, CBNRM in southern Africa can be interpreted as a new institutional model that links payments for ecological services to poverty reduction (Child 2004a,Frost and Bond 2008), which is applicable to many ecosystem services, but just happens to take advantage of the value of wildlife and opportunities that arose because wildlife provided an institutional greenfield. We certainly need such models. Ecosystem services have been valued at several times global gross domestic product (GDP) (Costanza et al. 1997), but we seldom take this into account in land use decisions. Indeed, the challenge of our generation may well be to internalize the costs and benefits of ecosystem services in the livelihoods and land use decisions of the rural people who coexist with biodiversity in the manner that has been attempted for wildlife. For example, the models of micro-governance and household cash payment systems developed in some community wildlife programs (Child 2004a,b,2006) are currently being replicated with REDD payments in Tanzania (Morgan Brown, personal communication). In other words, wildlife provides both a valuable resource for many drylands and also a new model with economic, institutional, and governance components that may be transferable to other natural resources and ecosystem services. This paper addresses only the economic component of this approach.

New Approaches to Wildlife Conservation

Beginning in the 1950s, conservationists in southern and East Africa began to develop a new paradigm for conservation based on sustainable use of natural resources (SASUSG 1996). Leading this policy experimentation was private game ranching, especially in southern Africa. Although this has been a highly successful conservation and economic initiative, relatively little is published about it or about the economic research, methods, and principles behind this growth. Private wildlife conservation or 'game ranching' strongly influenced the new vision of conservation and development that arose to prominence during the 1980s and 1990s (Martin 2009). The basic message of the sustainable use approach is that wild resources and ecosystem services are enormously valuable. If we change the ways that we govern them and if we can place landholders and rural communities at the junction of benefit and management, wild resources can pay for themselves and simultaneously address rural poverty and environmental injustice. At the time these ideas were introduced, they were radical, encompassing three major conceptual strands (Hulme and Murphree 2001):

That the state should devolve proprietorship, including the responsibility for and benefits from managing wild resources, to the landholders and the communities that live with them;

That natural resources should be exploited sustainably and as profitably as possible to achieve both conservation and development goals; and

That the neo-liberal concepts of markets, property, and exchange should play a greater role in shaping incentives for conservation and allocating resources to their highest valued uses.

Game ranching provides an example of how practitioners simplified a plethora of principles into one overarching guiding statement: 'to maximize the benefits from wildlife to the people on whose land it lives' (Child 1995). It is also an example of the centrality of institutions or the formal rules or informal norms that frame human conduct (North 1990) to the relationships between people, economy and nature, and the process of modifying these institutions in ways that were insightful, carefully crafted (Borgerhoff Mulder and Coppolillo 2005), and changed the outlook for wildlife in Africa.

Brief History of Wildlife Use

We can view the wildlife economy in four phases (Table 1). In the pre-modern economy, wildlife was plentiful, and the primary constraint to use was the technology and costs of harvesting it. This was followed by a 'frontier economy' associated with the Industrial Revolution and the exploration and settlement of Africa by Europeans. New technology and markets, including guns, wagons, and even railways, radically altered the economy of wildlife and enabled hunters to harvest vast numbers of wildlife at a low cost to sell in new urban and global markets. In the absence of institutions for controlling offtake, wildlife, including North America's vast herds of bison, was decimated in a classic case of market failure whereby individuals internalize benefits but externalize costs to society.

In response to this, wildlife was nationalized. In 1900 and 1933, the European powers met in London to respond to the perceived extirpation of wildlife in their African colonies, making three policy decisions with long-lasting implications (Heijnsbergen 1997). First, they encouraged the formation of state protected areas, which led to the formation of Africa's spectacular national parks and now support vibrant economies when managed effectively. Second, they greatly restricted the commercial use of wildlife, rendering wildlife valueless except for low-value subsistence uses. Third, they centralized ownership of wildlife in the state, disenfranchising landholders and taking upon themselves the burden of protecting wildlife from people (anti-poaching) and protecting people from wildlife (problem animal control). In some countries, local communities were severely disenfranchised, especially where traditional methods of hunting were made illegal and ownership of firearms was restricted (Carruthers 1989; Brockington and Igoe 2006). But in others, like Zambia and Botswana, local people were given considerable freedom to hunt under permit systems that often became difficult to administer (Astle 1999).

The policy response embedded in the 'London Convention' became widely accepted, sometimes dogmatically, as the way conservation should be done: a combination of 'pristine' parks and non-utilization of wildlife on land outside them, with wildlife as a priceless but commercially valueless asset, funded by the state and managed on its behalf by a small band of dedicated game rangers. However, by the late 1950s, population growth and the rapid spread of agriculture and livestock were causing an inexorable decline of Africa's unique and spectacular wildlife. At the twilight of colonial Africa, many leading conservationists met at the Arusha Conference on the 'Conservation of Nature and Natural Resources in Modern African States', emphasizing that a radical new approach was needed to conserve wildlife (IUCN 1963). This, the delegates said, needed to be led by Africans, and wildlife needed to become an economic asset according to the emerging 'use it or lose it' philosophy. Thus, the opening comments of the conference proceedings emphasize:

> Only by the planned utilization of wildlife as a renewable natural resources, either for protein or as a recreational attraction, can its conservation and development be economically justified in competition with agriculture, stock ranching and other forms of land use (p19, IUCN 1963).

Table 1: Phases in the political economy of wildlife

Phases in the political economy of wildlife	
Phase	Features
1. Pre-modern economy	Use is limited by ability/costs of harvesting
2. Frontier economy	Costs of harvesting greatly reduced by technology.
	Profits increased by markets.
	But few rules or norms to control use
3. Wildlife is nationalized	Control of wildlife centralized in the state
	Commercial use greatly restricted
4. Sustainable use approach	Use of wildlife devolved to landholders (and later, to communities)
	Commercial uses encouraged

The Emergence of Wildlife Utilization

Conservationists began to argue that wildlife was better suited to using Africa's harsh environments than the domestic monocultures imported from Europe because, to quote again from the Arusha Conference, the variety of ungulate fauna, their mobility, high standing biomass, and greater nutritional efficiency is seen to give them advantages over domestic livestock that concentrate their attention on a single constituent of the plant standing-crop biomass - the graminacious carpet (p19, IUCN 1963). Nonetheless, by 1980 fully, some 95% of the large herbivore biomass in southern Africa was livestock (Cumming and Bond 1991).

The 1950s and 1960s saw unprecedented research into the ecology of wildlife, its meat production potential, and even domestication (Talbot et al. 1961; Talbot et al. 1965; Mossman 1975). Cropping schemes were initiated, mostly in East Africa, and well-known conservationists like George Adamson and others (in Kenya), and Norman Carr (Zambia) proposed schemes whereby local people would benefit from wildlife (IUCN 1963; Parker 2004). The United Nation's African Special Project was developed to assess and invest in the potential for wildlife as an economic tool in Africa (Riney and Hill 1967), including the establishment of national parks and game reserves (e.g., Botswana) and cropping schemes (e.g., Luangwa Valley, Zambia). However, with the transition from colonial to African rule in East Africa, the control of wildlife was increasingly centralized (Kabiri 2010), and policy momentum shifted to southern Africa where there was considerable discussion about the potential value of wildlife (Riney 1960; Dasmann 1964), and experimental game cropping was initiated (Dasmann and Mossman 1961). By the late 1960s, the heads of wildlife agencies in southern Africa began meeting annually through the 'Standing Committee for Nature Conservation and the Management and Use of Wildlife' of the 'Southern African Regional Commission for the Conservation and Utilization of the Soil'. Soon thereafter, we see legislation emerging in Namibia, Zimbabwe, Mozambique, Botswana, and South Africa's provinces allowing landholders to use their wildlife commercially and with far fewer state-imposed restrictions (Child 1971; Suich and Child 2009). When it became apparent that permits and surveys were impractical and that landholders husbanded valuable wildlife just as they husbanded their domestic stock, regulatory requirements were reduced. In Zimbabwe, they were virtually removed by the bold Parks and Wild Life Act of 1975.

Ecology and Commodity (Meat) Production

Early research and discussions focused on wildlife's ecological adaptations to Africa's climate and diseases (Mossman 1975) and the ability of wildlife to produce more meat, or better quality meat, than livestock (Talbot et al. 1965). It suggested that wildlife produced high-quality meat but that the logistical challenges of cropping free-ranging wild animals reduced potential profits. Research then turned to the relative suitability of diverse multi-species production systems compared to livestock monocultures. In elegant studies, ecologists noted how the diversity of wildlife was matched to the diversity of vegetation and threats from predators (Jarman 1973), and that semiarid savannas in particular evolved under indigenous multi-species systems typically carrying 15 to 25 ungulate species compared to a maximum of five domestic species (Cumming 1982;Cumming 1995). Scientists were surprised at how much overlap there was in the dietary preferences of wild herbivores, but it was also argued that diet separation and the fact that gregarious wildlife was invariably mobile helped to maintain ecological balance, especially between the woody and grass components of savannas (Taylor and Walker 1978; Child 1995). Ecologists noted that production in many African rangelands was water limited (Coe et al. 1976; Bell 1982) and emphasized the importance of the grass layer in absorbing rainfall and making it available to plants (Kelly and Walker 1976). 'Degradation' was thought to be synonymous with the loss of perennial grasses, which maintained the structure of the soil surface and absorbed water (Riney 1963; Walker 1987). Overgrazing or frequent incorrect burning of the grass layer could tip the crucial balance between grass and trees (Riney 1963; Mahesh Sankaran et al. 2005) into a new (and less productive) stable state from which recovery was difficult (Campbell and Child 1971;Walker et al. 2004).

Important and interesting as this research was, researchers were marching down a blind alley in comparing the relative ability of wildlife and livestock to produce meat. The meat productive potential of semiarid rangelands was ecologically limited, regardless of species. Increasing animal production reduced the health of the grass layer (and water infiltration where water was the limiting factor), thus undermining itself. In other words, meat production was ecologically limited, and attempts to move beyond these limits ultimately failed as we see later with private beef ranching in Zimbabwe. Ultimately, the solution lay not in choosing between wildlife and livestock for meat production but in adding value in ways that was not directly linked to extraction from the environment, such as experiences or luxury goods like trophy animals. Seen retrospectively, wildlife scientists mistakenly thought in terms of commodity production (like the agricultural agencies of the time) rather than improving the economic allocation of resources to add value. In the end, institutional adaptation and the development of new markets by farmers, not production technology, provided commercial solutions for ranchers through a shift from agro-extractive beef commodity production to a bio-experience economy.

Dysfunctional Legal Systems and Human-Wildlife Relationships

In the twentieth century, white settlers acquired land in southern and East Africa, and African agriculture and populations grew rapidly. Prior to World War II (WWII), the new conservation laws were directed mainly at white commercial hunters, and local hunting for the pot was usually overlooked (Parker 2004). But as rural populations grew, local hunting began to threaten wildlife, was increasingly treated as poaching, and prosecuted by the state. Criminalizing a traditional livelihood strategy like hunting created a powerful sense of injustice, especially when enacted by colonial governments that lacked political legitimacy. 'Poachers' became Robin Hoods and were protected by their communities, so wildlife laws were difficult to implement. This also created a negative relationship between local people, wildlife, and the state.

Setting the state up against the people was never likely to result in sustainable solutions for wildlife. In pre-independence Zimbabwe, for instance, protectionist conservation alienated white farmers from wildlife and became an important political tool in the hands of black nationalists by highlighting the injustice of white minority rule. Even today, some national wildlife agencies expropriate the benefits of wildlife for themselves and their cronies and treat local 'poachers' brutally, so pastoral people who are natural allies of wildlife, are planting fields in important wildlife areas, mostly to protect their land rights (Nelson 2010).

One outcome of these dysfunctional policies is the false impression that local people hate wildlife. Our surveys of local people across southern Africa show that people are often extremely concerned about the impact of wildlife on their crops and livestock (and the ineffective management of this problem), yet over 90% of people claim to like wildlife (or to like it a lot) for both aesthetic and material reasons. What they do not like, however, is the inequity of wildlife laws and the high-handed manners of officials who implement them; they like wildlife, and even protected areas, but not the wildlife officials or laws. This situation is beginning to improve with the advent of CBNRM in countries like Namibia and Botswana.

Protectionist, centralized wildlife legislation, from urbanized and wealthy industrial nations, was politically unjust and economically flawed, causing enormous losses of wildlife in developing, rural societies after WWII as agriculture expanded. The 1940s to the 1970s were the heyday of agriculture. Cattle ranchers complained that 'you cannot farm in a zoo' and wildlife was steadily exterminated (Taylor 2002). Newspaper clippings in Zimbabwe dramatize cases where cattle ranchers shot large numbers of wildebeest and zebra and (having no legal commercial use) left them to 'rot in the veld'. However, the bigger threat was more subtle: the fencing of water points and livestock overgrazing eliminated wildlife for lack of food and water.

The Perspectives and Contributions of Private Landholders

We use a series of case studies to describe the shift back to wildlife, which began slowly led by a few maverick cattle ranches and then accelerated into a landscape-wide phenomena. In the Lowveld of Zimbabwe, for example, the late George Style of Buffalo Range Ranch (20,000 ha) had a penchant for wildlife. From the mid-1950s, he developed a successful cattle business, fencing paddocks and providing artificial water. The eastern section of his ranch was bounded by the Chiredzi River, which for decades provided wildlife and livestock with permanent water in the dry season (Figure 1). Consequently, this part of the ranch (some 9,000 ha) was degraded, and George Style experimented with wildlife as much as a hobby as for commercial reasons. Browsers like eland, kudu, and impala thrived, and the sensitive grazing species like sable, roan, and Lichtenstein's hartebeest displaced by cattle began to recover. Style culled impala to be sold through his butcheries and soon (early 1970s) initiated 'mini-safaris' with foreign hunting clients in a well-appointed hunting camp.

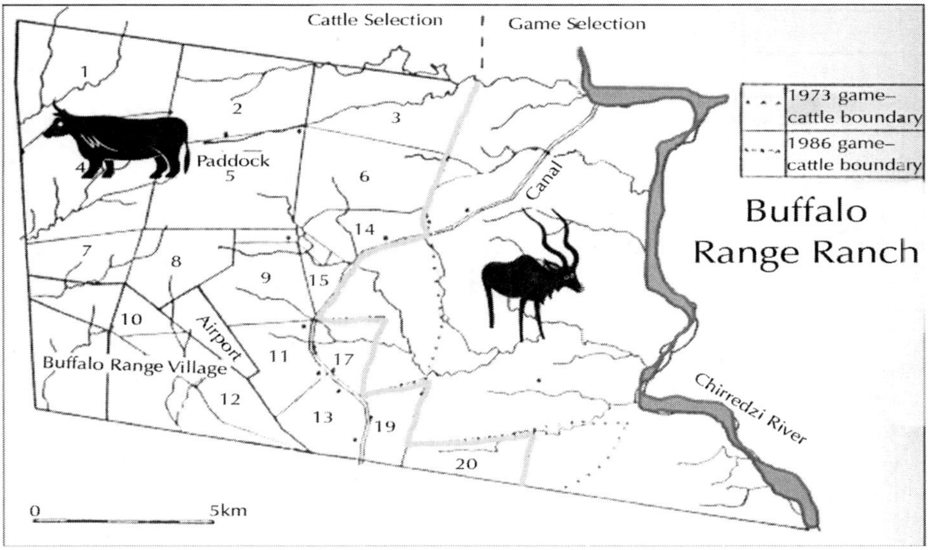

Figure 1: Map of Buffalo Range Cattle and Game Ranch.

George and his son, Clive, encouraged the use of their ranch for comparative research. Detailed vegetation transects showed that cattle grazing damaged perennial grasses and the soil surface, largely because of the continuously high stocking necessary to keep a beef enterprise viable, whereas the range was slowly recovering under wildlife (Taylor and Walker 1978; Child 1988) (Figure 2). Overstocking appeared to directly damage range productivity, with the pattern of cumulative overstocking (i.e., stocking rate in each year less predicted carrying capacity calculated from rainfall using the formula from Coe et al. 1976) matching a rapid decline in cattle calving rates (Child 1988) (Figure 3) and ultimately destroying the profitability of the livestock enterprise on Buffalo Range. This experience was shared by many cattle ranchers in the Lowveld (PriceWaterhouse 1994; Taylor 2002), who began to switch to wildlife enterprises, some partially, some completely, and some as groups of landholders over large blocks of land, especially after losing livestock in the severe droughts of 1984 and 1992 (Taylor 2002; Lindsay et al. 2009).

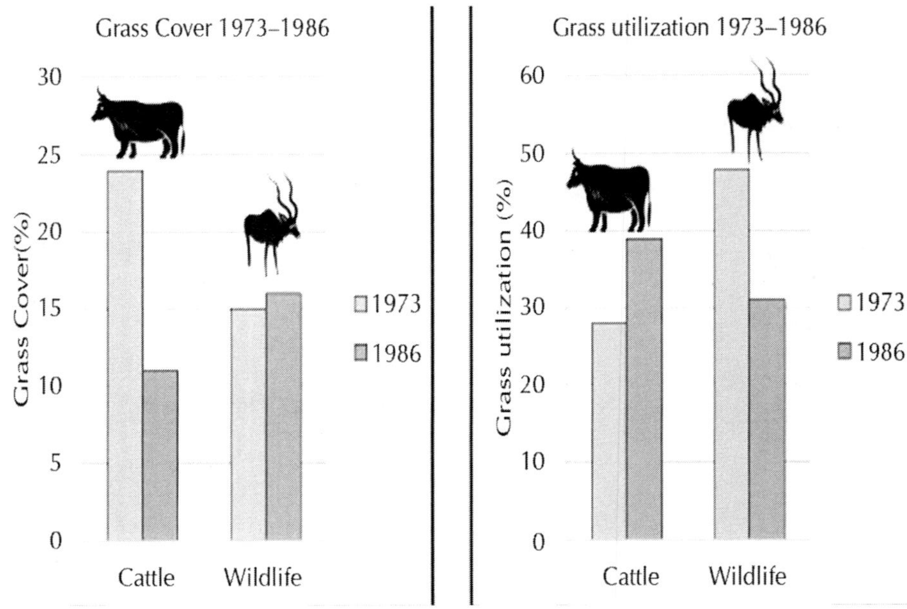

Figure 2: Changes in grass cover and grazing intensity on Buffalo Range ranch between 1973 and 1986.

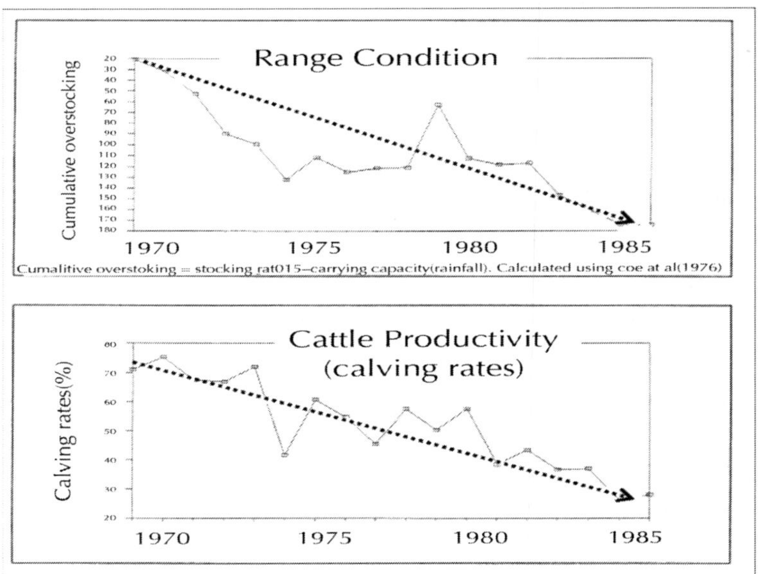

Figure 3: The relationship between range condition and cattle productivity on Buffalo Range ranch (1967–1986).

The Beginnings of Safari Hunting on Private Land

On most properties, the switch to wildlife was gradual. Many cattle ranchers survived financially by using remnant populations of wildlife for safari hunting. With good revenues and few overhead costs, wildlife initially supplemented cattle ranching, but later, some ranchers switched over entirely. Hunting guests from North America or Europe stayed with the rancher for seven to ten days, initially in a spare bedroom and later in specially built accommodation. Each day, they would go out early in the morning or in the late afternoon, shooting roughly one animal a day. These were called 'plains-game safaris' because they relied on species like kudu, zebra, wildebeest, impala, and warthog. Ranchers then improved their hunting product by nurturing or reintroducing valuable species like eland, sable, waterbuck, and even predators like leopard, and later by adding big game like elephant and buffalo, usually as part of multi-property conservancies. But to do this, they began to improve environmental management by reducing stocking rates and habitat fragmentation by removing cattle fences and scaling up to multi-property management units. Cattle ranchers like George Style spent many days in the field hunting and developed considerable knowledge about the habits and needs of wildlife, which spread through farmer field days. Under natural conditions, wildlife is highly mobile to respond to highly variable savanna rainfall, and ranchers defragmented their land with simple (and cost-saving) measure like removing one strand from conventional four-strand cattle fences, allowing species like kudu and eland to jump over them, and wildebeest and warthog to duck under them. Similarly, they stopped fencing water points and modified conventional cattle water troughs to enable wildlife to drink. They catered for the specific needs of valuable species like sable, which struggle to compete with cattle, and managed diseases like 'snotsiekte' (malignant catarrh) by ensuring that their cattle were separated from wildebeest at calving time, during which transmission from wildebeest to cattle occurs.

Trophy Hunting is Sustainable; Meat Production Very Risky

Although it runs counter to the moral argument, hunting animals for meat should not be encouraged whereas killing them for sport provides a powerful conservation tool. The recovery of wildlife over much of southern Africa has gone hand-in-hand with trophy hunting, which is a robust, low-risk solution for two reasons: (1) the mathematics of population biology and (2) the transparency of safari hunting markets. An offtake of 2% to 3% provides a steady supply of quality trophy males compared to population growth rates of 10% to 30%, so ranchers like George Style made good profits from safari hunting at the same time when wildlife populations recovered rapidly. Trophy hunting (unlike meat production) is profitable and ecologically robust, even more so because trophy hunting markets are personalized and information-rich, so hunting clients quickly avoid outfitters who over-hunt with below-average trophy (horn) size. Indeed, the ecological challenge on successful game ranches like Buffalo Range is seldom too little wildlife, but too much. George Style culled hundreds of impala to protect the environment and allow greater diversity of wildlife even though game meat production seldom covered costs.

In contrast to safari hunting, which is ecologically robust and profitable, commercial or subsistence wildlife meat production is ecologically risky and economically questionable. Meat is a ubiquitous low-value commodity without the market checks and balances associated with hunting trophies. We need to kill at least 20 times as many animals to generate the same income from meat as from trophy hunting, and the costs are much higher. On private land, bushmeat is seldom viable except as a by-product of more lucrative hunting and tourism businesses, with the implication that bushmeat poaching is viable only because the harvester externalizes many costs to society. The safest conclusion is that meat production in drylands, either from livestock or wildlife, is seldom viable if the full financial and environmental costs and benefits are accounted for. It has not succeeded on private land, and we extend economic models based on bushmeat production to rural communities at our peril.

Developing an Economic Understanding of Wildlife Production

Trophy hunting is worth more than a billion collars in South Africa alone (Flack, personal communication), yet data and publications on private wildlife conservation in southern Africa are sparse. The purposes of this section are to summarize, with increasing sophistication, methods used to understand the economics of the private wildlife sector, to influence policy making in the past 25 years, and to use case studies to illustrate how the wildlife sector has emerged.

Shortly after Zimbabwe's Independence in 1980, one of the authors (Brian Child) was recruited by the wildlife department to conduct economic research and extension on game ranching (1984 to 1989). By spending extensive time with game and cattle ranchers, he gained their confidence, learned how much they knew about wildlife, and worked with them to understand the economics of their wildlife and livestock enterprises. Data from 22 cattle ranchers at the wetter limits of rangelands in the Zimbabwean Midlands in 1984 (Vincent and Thomas 1961) showed that safari hunting was easily the primary wildlife activity, with income comprised as follows: guiding (35%), trophy fees (46%), meat from trophy animals (13%), and meat cropping (7%). All but one of these landholders was primarily a cattle rancher.

Gross Income (Turnover) and Net Margins (Profits)

Livestock, which comprised 80% of herbivore biomass, generated most of the income in the area (Table 2), but when variable costs were deducted, livestock profits were (Zimbabwean dollar) Z$4.52/ha compared to Z$2.93 from wildlife. All values for Zimbabwean research are stated in 1984/86 Z$ when US$1=Z$1.3 A new financial ratio (i.e., profit/kg livemass) that we invented to reflect the ecological reality that grass was the limiting factor (see above) showed that, from 1 kg of livemass, wildlife earned 17c and cattle earned 7c, suggesting that, at the margin, it paid landholders to reallocate grazing from livestock with wildlife. A similar survey of 15 properties and 446,818 ha in the more arid Lowveld in 1986 (including Buffalo Range Ranch) confirmed that it was economically rational to switch land from livestock (earning 7c/kg livemass) to wildlife (26c) (Table3).

Table 2: A financial comparison (in Zimbabwean $ dollars) of wildlife and livestock in the Zimbabwean Midlands in 1984

	Units	Cattle	[redacted]
Comparison/unit area (wildlife = 20% of biomass)			
Income	Ha	$13.57	$3.36
Profit (income – variable costs)	Ha	$4.52	$2.93
Comparison/kg biomass			
Gross income	Kg	20c	20c
Profit	Kg	7c	17c

Table 3: A financial comparison (in Z$ dollars) of wildlife and livestock in the Zimbabwe Lowveld in 1986

	Livemass (kg)	Turnover (Z$)	Profit (Z$)	[redacted]
Cattle	10.1 m	$2.0 m	$0.7 m	0.07
Wildlife	4.7 m	$1.4 m	$1.2 m	0.26

m = million.

Child *et al.*

Repeating a survey of rancher opinions (unpublished mimeo) about rancher opinions about wildlife in the Zimbabwe Midlands ten years after the radical Parks and Wild Life Act of 1975 (that devolved full authority to landholders to use wildlife), and only four years after the civil war ended, showed significant changes. Wildlife income had quadrupled, and neutral or negative attitudes towards wildlife decreased from 22% to 0%. In line with the predictions of the sustainable use approach, the geographic range of valuable hunting species increased, as we also see in the data from South Africa (see below). The number of properties utilizing reedbuck, bushbuck, wildebeest, zebra, waterbuck, tsessebe, kudu, eland, and sable increased by 21%; the geographic range of these species expanded by 22%, and that of the most valuable species (i.e., waterbuck, tsessebe, eland, and sable), by 35%. In terms of biodiversity conservation, a single, sedentary, gregarious bulk-roughage feeder (i.e., domestic cattle) was replaced by a spectrum of 15 or more species of wild (and domestic) animals (Child 2009), and habitats noticeably improved (personal observation, photographs) on large unfenced ranches in Zimbabwe, although there were strong concerns about degradation on small fenced game farms in South Africa.

By the late 1980s, most cattle ranchers in Zimbabwe began to promote and use wildlife because of a decline in cattle prices and subsidies and increasing ecological problems highlighted by the 1984 and 1992 drought. A significant number shifted completely into wildlife, especially in southern Zimbabwe and near Hwange National Park and Victoria Falls. An excellent documentary video about the formation of the Save Valley Conservancy illustrates these broader trends (Taylor 2002). In the video, one of the leading ranchers in the Lowveld (Clive Stockil) describes how cattle ranching had been developed in the areas since the 1950s. Following the adage 'you can't farm in a zoo', wildlife had been steadily replaced, including through active elimination campaigns (e.g., of buffalo) by the veterinary department. Footage shows the degradation of the habitat (much as described for Buffalo Range) and the collapse of cattle enterprises. Noting that ranching needed to be ecologically sustainable, economically viable, and socio-politically acceptable, Stockil describes how 27 ranchers studied the situation. They commissioned a report (PriceWaterhouse 1994) that concluded that wildlife would double their income and increase the return on capital from 1% to 3% (beef) to 10% to 22% (wildlife). Consequently, the 27 landholders in the Save Valley shifted from livestock to wildlife, combining 344,200 ha of previously degraded cattle ranches into a huge wildlife conservancy with no internal fencing and common management policies (Lindsay et al. 2009). The Save Conservancy was systematically restocked with elephant, buffalo, and predators including lions and wild dogs, and many of the antelope species displaced by livestock. Hunting and tourism lodges diversified the economy, generated more profits, provided more jobs, and allowed the environment to recover. To all intents and purposes, this resembled a private national park. At least seven other similar initiatives occurred in Zimbabwe at the time.

Kickback from Single Commodity Leviathans

By the late 1980s, the relative economics of wildlife and livestock had reversed, but agricultural polices had not. Despite falling beef prices on the world market, government veterinary and agricultural agencies continued to support beef while imposing considerable costs on the wildlife sector to the detriment of private ranches and the greater economy (Muir-Leresche and Nelson 2000) although wildlife was enabling cattle ranchers to stay on the land (a pattern also observed in Texas, personal observation) and to recover its ecological function (Child 1988; Jansen et al. 1992;PriceWaterhouse 1994). But powerful government agricultural agencies continued to promote beef without considering its impact on wildlife, illustrating the dangers of single-sector decision-making. Between 1919 and the 1980s, thousands of wild animals including rhinos were shot in the name of tsetse fly control (Child and Riney 1987), hundreds of miles of game fence were erected, habitats were grossly modified including the bulldozing of riparian woodlands and waterholes, and DDT and deildrin were applied on a broad scale. In the 1980s, large areas of the country were further fenced to control foot-and-mouth disease despite questions as to the technical and economic merits of a fencing solution (Taylor and Martin 1987). Based on controversial evidence of buffalo-to-livestock transmission, all buffalo on private land were shot despite the veterinary department being informed that buffalo could double the profitability of wildlife enterprises. To combat this, the small wildlife department began to study the economics of the cattle sector and developed a quick and dirty 'budget' for the beef sector (Table 4). Financial data compiled from government reports (Child 1988) showed that even preferential access to the high-priced European Union market would not cover farmer and industry costs, even before the fixed costs to farmers (e.g., capital value of livestock, land, and facilities), the industry (e.g., costs of new EU compliant abattoirs), or the environment were accounted for. Single-sector Leviathans emphasized commodity production (e.g., bags of maize and kilograms of beef) with little appreciation of economic issues like comparative advantage or allocative efficiency, which may speak well to politicians but does not maximize societal well-being.

Table 4: Quick and dirty beef industry budget (Zimbabwean $ dollar 1986)

Costs	
1. Government services	62
2. Cold Storage Commission subsidy	50
3. Farmer's variable costs	74
4. Farmer capital investment (land, cattle)	??
5. Farmer overhead costs (fixed costs)	??
6. Industry investment in European Union-standard abattoirs, buffalo fencing, etc.	??
7. Environmental costs (overgrazing)	??
Total Costs	185+
Income	
1. Actual	149
2. With European Economic Commission exports	191

A historical review of the government's own reports, including Parliamentary Commissions of Inquiry, showed that the beef sector had been heavily subsidized for several decades through advantageous beef prices, wealth transfer from the peasant to commercial herd, support to farm development through fencing and water provision, subsidisation of single-channel marketing systems (the Cold Storage Commission), and so on (Phimister 1978; Child 1988). Methodologically, developing these arguments is time consuming and subject to different interpretations. Parity pricing provides a quicker and more reliable method for assessing if beef producers are subsidised (or taxed) by comparing the price farmers get for beef on the domestic market compared to the world market. Using government statistics (Figure 4) reveals that, after the mid-1970s, the domestic price of beef was on average 40% higher than the world price, confirming other research findings that producers were heavily subsidised through pricing policies (Williams 1993).

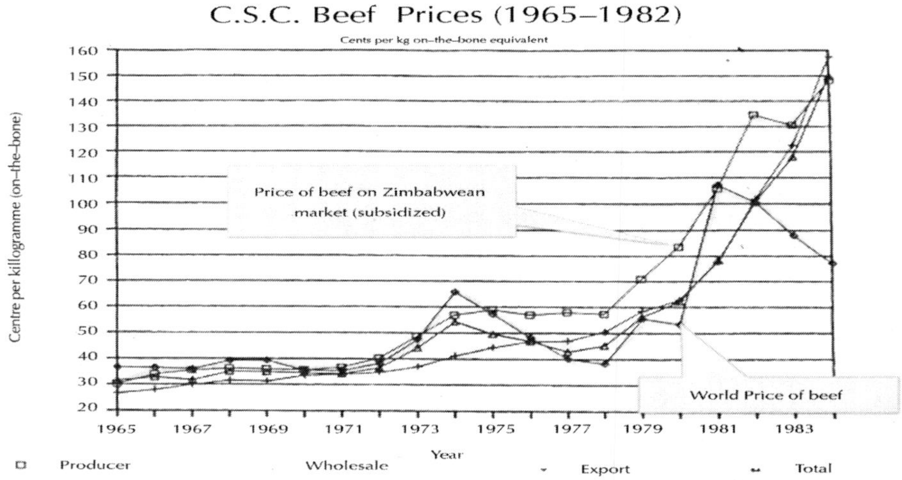

Figure 4: Comparing domestic and world prices to estimate if land use is taxed or subsidized.

Financial and Economic Analysis

The proponents of wildlife intuited as far back as the 1960s that wildlife was being displaced by livestock despite being a better land use option. An economist would say that wildlife had an inherent comparative advantage in drylands but was undervalued by market failures so that scarce resources were being misallocated to the detriment of society. In 1990, World Wide Fund for Nature (WWF) and Zimbabwe's wildlife agency used detailed financial accounts and asset registers to assess the viability of some 150 cattle and wildlife enterprises (Jansen et al. 1992). Wildlife was more profitable than livestock, with only 5% of livestock operations generating a return on capital in excess of 10% when profits were calculated using market (financial) prices (Figure 5). Market prices were then adjusted to remove pricing distortions caused by market failures. In terms of societal outcomes, wildlife was even more profitable (Figure 6), providing jobs, economic growth, and foreign currency (which were undervalued by government-controlled exchange rates). The rapid transition towards wildlife in many areas, unfortunately, was never accurately quantified, but judging from trophy sales, wildlife populations on private land quadrupled between 1984 and 1992 (Booth 2002).

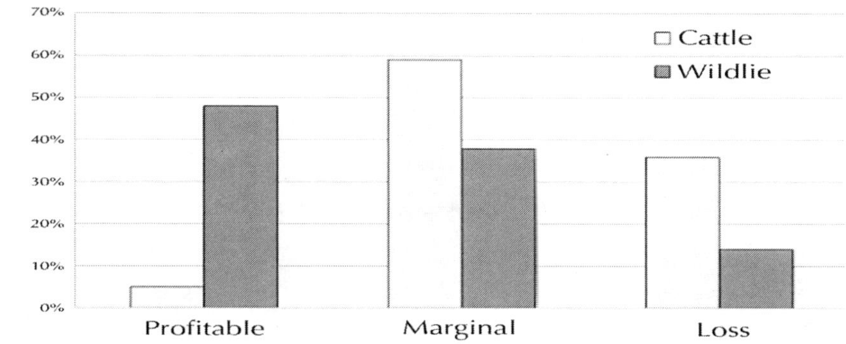

Figure 5: Comparison of the financial profitability of cattle and wildlife in Zimbabwe in 1990.

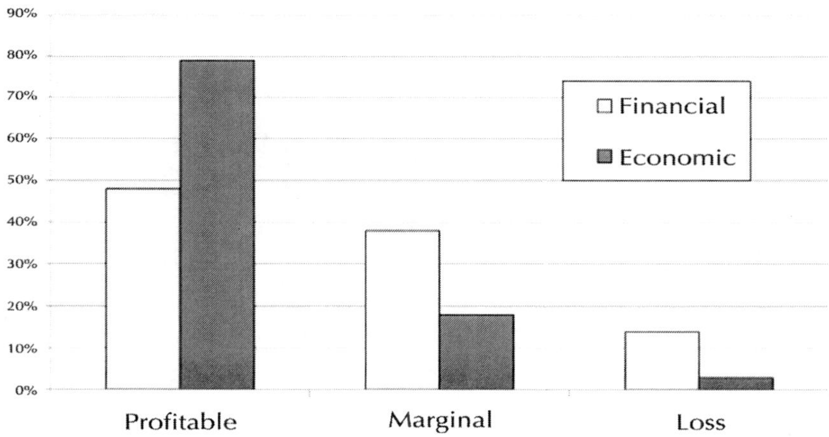

Figure 6: Comparison of the financial and economic profitability of wildlife in Zimbabwe in 1990.

South Africa and Namibia

The land invasions in Zimbabwe in 2000 undermined much, but not all, of this progress. However, similar trends on private land in Namibia, South Africa, and Botswana confirmed the coming of age of the wildlife bio-experience sector. In Namibia, between the early 1970s and 2001, livestock populations on private land halved from 1.8 million to 0.91

million, whereas 'huntable' wildlife populations doubled from 0.565 million to 1,161 million head for economic reasons (Barnes and Jones 2009). In South Africa, the numbers of wild animals easily quadrupled between 1964 and 2005, with some sources (Flack unpublished data; Carruthers 2008 suggesting as many as 18.6 million wild animals between 1964 and 2005, and there are now some 10,000 wildlife ranches and some 4,000 mixed wildlife and livestock enterprises (Bothma et al. 2009). Private wildlife areas expanded eightfold between 1979 and the mid-1990s (Chadwick 1996), protecting at least 16.8% of the total area of South Africa (Cousins et al. 2008) compared to 6% as IUCN category I to IV protected areas. As with Zimbabwe, 'utilized' species increased much more rapidly than those 'protected'. Rhino increased from 50 to 18,000, and black wildebeest from a few to 26,000, much faster than bontebok (few to less than 5,000) and Cape Mountain Zebra (few to 1,200) for which hunting has only been allowed within the past decade (Flack, personal communication).

Policy Analysis Matrix

Despite its size and economic importance, there is remarkably little research or political economic understanding of the wildlife sector in South Africa and some risk that conservation, economic, and employment gains will be undone in the politically charged debate over land restitution. We highlight key findings of an investigation of the economics of game ranching near Kruger National Park (Musengezi 2010) and illustrate the usefulness of the Policy Analysis Matrix or PAM methodology (Table 5).

Table 5: Policy Analysis Matrix (PAM)

		Costs		Profits
	Revenues	Tradable inputs	Domestic factors	
Private prices	A	B	C	D
'Social' (or 'economic') pricesi	E	F	G	H
Divergence	I	J	K	L

[1] The PAM can also account for environmental externalities and for ecological, regulating and social/cultural services by building estimates of the cost of these into social prices, but we did not do this.

The PAM is theoretically robust (Monke and Pearson 1989), relatively easy to apply, and useful for influencing policy makers, especially in finance ministries because it is easy to understand as it follows the basic calculation, profit = income from sales less costs of production. Constructing a PAM requires social skills because it relies on landholders' providing their income and expenditure accounts. It also requires technical skills to classify accounts into revenues, tradable and domestic factor costs, and to calculate 'economic' prices. It can be extended to incorporate negative externalities on ecosystems or social/equity systems and to guard against ignoring these in policy recommendations. The PAM is particularly useful for assessing if a land use represents a good choice for society and how much choices are being distorted by policy and market failures.

In this paper, for example, we use PAM to assess if wildlife is an optimal use of drylands, that is, does wildlife have a comparative advantage? 'Comparative advantage' is not the same as absolute advantage but explains why countries gain mutually from trading in products for which they have relative efficiencies. Comparative advantage is calculated by comparing value-added (i.e., revenues-tradable inputs) to the costs of local land, labour, and capital (i.e., domestic factors), which is why the PAM split costs into two columns: tradable inputs and domestic factors.

Second, we can calculate how much market and policy failures are distorting prices, and therefore land use decisions, away from the real values that would guide private actors to allocate resources efficiently for society. We quantify these price distortions by comparing private prices (top row) to prices in perfect markets (social prices row), correcting prices using various econometric techniques. The divergence between the private and social prices is a measure of the magnitude of market failure. It is also a critical economic concept with considerable implications for conservation policy. When private prices diverge from their true values, private decision makers do not properly internalize social benefits and costs of their actions. This causes under-provision or over-utilization of natural resources and is the root of many natural resource problems (Libecap 2009).

In 2010, Jessica Musengezi applied the PAM methodology to eight properties near Kruger National Park in South Africa (Table 6). Wildlife enterprises were too diverse (and innovative) to summarize using simple averages. Property (a), for example, supported high-end tourism charging over US$500 per night, while property (d) was purchased for ecological reasons and profits were an insignificant objective. Only property (g) still had livestock, but profitability was achieved from vertical integration including an abattoir and butcheries (providing profits far in excess of cattle ranching in the area).

Table 6: Summary of PAM results for eight wildlife properties in South Africa in 2010

Farm								
All price in South Africa Rand (US $1=R7)	a	b	c	d	e	f	g	h
Property attributes								
Farm size (ha)	5,207	1,700	3,700	2,017	2,800	3,200	14,400	30,000
Big five	Yes	No	Yes	No	No	No	No	Yes
Beds	68	42	52	12	12	10	-	-
Average price (rand)	2,854	350	260	285	350	250	-	-
Revenue sources(% of total revenue)								
Tourism accommodation	95	2	79	95	0	5	0	0
Biltong hunting	0	0	0	0	0	17	1	0
Cattle	0	0	0	0	0	0	84	0
Game meat	0	0	0	0	0	1	0	0
Live game sales	0	92	2	0	100	71	14	100
Retail	6	0	3	2	0	0	0	0
Trophy hunting	0	6	16	0	0	6	2	0
Labour								
Number of workers	194	84	20	19	9	10	75	10
Labour cost (rand/ha)	2,788	600	664	471	238	87	233	14
Labour cost (% of total expenditure)	77	21	43	63	51	18	19	18
Financial and economic results								

Gross income (rand/ha)	8,706	4,936	2,231	105	1,643	1,099	2,906	150
Gross margin (rand/ha)	8,282	2,886	856	4	1,434	768	1,947	90
Private cost ratio	0.38	0.44	0.13	−1.51	0.34	0.69	0.16	0.16
Domestic resource cost	0.2	0.22	0.59	−1	0.26	0.43	0.11	0.08
Effective protection coefficient	0.94	0.88	0.92	0.94	0.96	1.03	0.91	0.9
Profitability coefficient	0.73	0.64	0.14	1.18	0.86	0.55	0.86	0.83
Net policy transfer (% tax on net profits)	27	36	86	−18	14	45	14	17

PAM, Policy Analysis Matrix; N/a.

Simple gross income and net margin (i.e., gross income − direct costs) data showed that all wildlife properties (except (d)) are profitable (Table 6), exceed the returns from extensive commercial beef production (according to models provided by the Department of Agriculture), and provided significant numbers of jobs (Musengezi 2010). Indeed, wildlife enterprises create more jobs than the livestock enterprises they displaced by a factor of two and as much as five, and these jobs are also more specialized, increasing wage bills some 20- to 32-fold, (PriceWaterhouse 1994; Taylor 2002; Langholz and Kerley 2006).

We illustrate how the PAM is used to assess the relative economics of wildlife production using Farm (a). The capital letters in the formulas are defined in the PAM in Table 5. At a basic level, Farm (a) generates (South African rand) R8,706 /ha and is 'profitable' after costs with a gross margin of R8,282 (US$1=R7). It also adds value to the South African economy from both a private and social perspective. In terms of the prices that guide the farmer to make land use decisions, the private value-added exceeds costs of domestic factor inputs. Said technically, the private cost ratio shows that, for each rand of net profit (A-B), 38cents goes to pay domestic factors (C). This leaves a net profit of 62 cents which is substantial (i.e., PCO = C/(A-B) = 0.38). When prices are adjusted to correct pricing distortions, wildlife generates one rand of profit from 20 cents of domestic factors costs (i.e., Domestic Resource Cost Ratio (DRC) = G/(E-F) = 0.196). Economically speaking, wildlife has a comparative advantage and is a good use of drylands in South Africa.

We can now use the PAM to advise policy makers as to how market failures are effectively taxing or subsidising wildlife. First, we assess the impacts of exchange rate and trade policies using the Effective Protection Coefficient (EPC). On Farm (a), the social value-added (E-F) exceeds the private value-added (A-B) by only 6% (i.e., EPC = (A-B)/(E-F) = 0.94), which confirms that product markets in South Africa are relatively free and only slightly disadvantaging wildlife. Second, we use the profitability coefficient to measure the overall effect of all policy distortions on the enterprise. On Farm (a), this is D/H = 0.73, so overall, Farm (a)'s private profit is 27% lower than it would be in the absence of economic distortion. This suggests that, in South Africa, domestic factors (i.e., immovable land, labour, and capital) are moderately distorted and that, if we corrected these issues, wildlife would be 27% more profitable.

There is considerable variability amongst the eight properties, depending upon the specific configuration of their enterprise and cost structures. Nonetheless, wildlife clearly provides significant jobs from marginal land. It is an economically efficient use of drylands in South Africa (DRC < 1). It is also profitable to farmers despite government policies being disadvantageous to wildlife. However, Musengezi's research also showed that game ranchers faced significantly more nonfinancial barriers compared to conventional agricultural enterprises in the form of additional regulations and permissions. Her analysis is a single snapshot in time and does not incorporate habitat recovery or the considerable increases in land values associated with wildlife properties.

Conceptualizing Wildlife as an Economic Option

We can use two conceptual diagrams to summarize the economics of wildlife. Figure 7hypothesizes that wildlife outcompetes livestock agriculture in some drylands because it is easily diversified through hunting and tourism. These advantages are magnified at the economic or societal level because the wildlife sector is associated with vertical integration (lodging, guiding, etc.) and tourism multipliers (e.g., backwards linkages to inputs like food, wine, transport, etc.) that are far higher than for simple commodities (like meat).

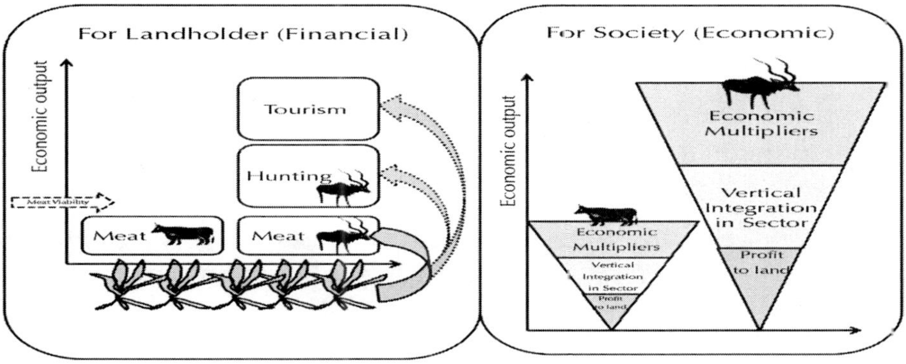

Figure 7: Conceptualizing economic differences between the bio-experience and commodity economy in dry lands.

This has important policy implications. In terms of development, wildlife utilization is a sensible land use option judged purely in terms of jobs and economic growth in some dry lands (Figure 8). Therefore, policies that internalize both the costs and benefits of wildlife are likely to result in positive conservation and development outcomes. Conservationists also need to take this seriously, recognizing that national parks in many savannas can achieve their conservation objectives while also functioning as engines for economic growth and rural development, which is also an important social responsibility.

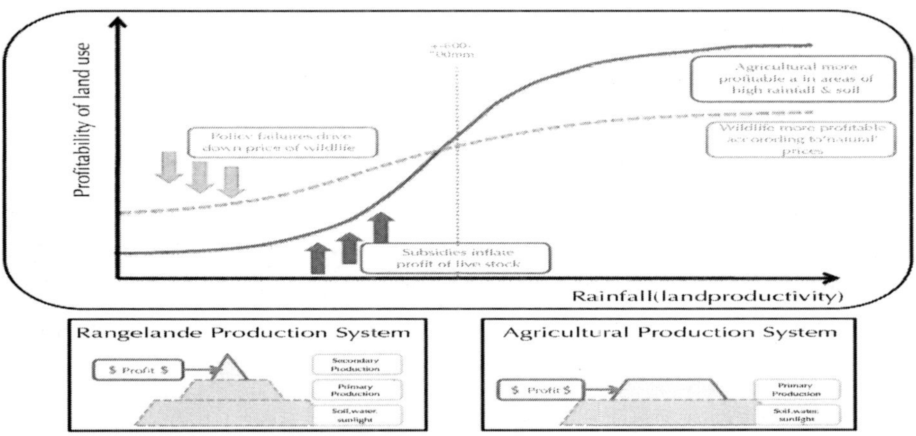

Figure 8: Conceptualizing production systems (bottom) and comparative advantage (top).

In higher-rainfall agricultural areas, the argument is reversed (Figure 8). If we want to conserve representative samples of these environments, we will need to subsidize them (e.g., protected areas) because wildlife is only economically competitive on patches of land that are difficult to farm (e.g. wetlands and hills) or uniquely accessible to markets in cities.

Figure 8 also illustrates a critical difference between agriculture and natural resource-based systems. In traditional agricultural zones, ecosystems are greatly simplified, and environmental energy is channeled directly into one, or at most a few, products like cotton or corn. Outside these zones, people tend to harvest plants indirectly using animals (e.g., drylands) or to harvest a much wider variety of plant and animal products (e.g., tropical forests). In drylands, humans harvest the third trophic tier. There is a quantum loss in productivity because energy is lost in transfers between trophic layers, and commodity production is ecologically limited (illustrated in Figure 8). It is in these three-tier environments that the bio-experience economy has a comparative economic advantage over agro-extractive commodity production. However, historically, the wildlife in these ecosystems was undervalued and replaced by livestock. Policy and market failures, especially the nationalization of wildlife, restrictions to commercial use, and subsidization of commodity sectors, sent out the wrong pricing signal to landholders. This trend was only reversed when radical changes in wildlife policy began to address market failures by (a) devolving use rights for wildlife to landholders and (b) encouraging commercial use. Where market failures remain pervasive, for example where wildlife remains nationalized (e.g., Kenya), and in communal lands where access to resources remains open, wildlife continues to be replaced (see below).

Institutions and Value

Walker et al. (2004) use the example of private conservation in southern Africa to illustrate how complex systems flip from one zone of attraction to another. Under early colonial policies, ranches in southern African was locked in a mutually reinforcing cycle of environmental degradation and financial crisis associated with beef commodity production, but when these policies were changed, landholders quickly adapted, giving rise to a dynamic and profitable wildlife sector. That the natural environment is the same but that radical change followed a change in the rules, alerts us to the importance of these rules and introduces the field of institutional economics (North 1990). Institutions are the rules and norms that guide human behaviour, including the allocation of resources. Institutions like property rights and markets allow wealth to be created through processes of specialization, diversification, and exchange, so institutionally rich countries are usually wealthier than institutionally poor ones (North 1995). We hypothesize that this same logic applies to drylands: institutionally rich drylands produce much greater value than those that are institutionally thin because richer institutions enable landholders to convert ecological raw materials into higher-order goods and services without damaging the resource base. Thus, new institutions like wildlife property rights, tourism markets, banks, wildlife auctions, etc. have added considerable value to wildlife through processes of specialization, diversification, and economies of scale, investment, and exchange. New wildlife institutions add value, allowing land to generate upward from South African Rand R1,000/ha, and some R100/ha, to R2,788/ha in wages alone (Table 6). Moreover, the devolution of wildlife rights to 10,000 private landholders has fueled considerable innovation in the development of wildlife products, which has driven up the value of wildlife in ways that have not happened when natural resources are monopolized by the state. Thus, southern Africa provides a multitude of hunting and tourism experiences from walking with lions to having moonlit dinners in the bush, whereas in Kenya and Tanzania, most tourists are restricted to closed vehicles in state-managed wildlife areas. We can use the price of South Africa's relatively unspectacular wildlife as a benchmark of its potential value. This suggests that the value of wildlife is far lower than it should be in centralized and institutionally thin wildlife sectors, including many communal lands.

Our discussion of the relationship between the institutional richness and economic productivity of drylands reflects Douglass North's explanation for why some countries are rich and others poor. North postulates that the most dramatic shift in human history has been from a personalized economy of big men and small-scale economic and political activity to an impersonal economy that relies on rules that allow us to transact reliably all over the world and with people we don't know, allowing humans to specialize and diversify on much higher scales than ever before (North 2003).

In Zimbabwe, enriching wildlife institutions by devolving proprietorship to landholders, encouraging trade, and establishing institutions like producer associations and auctions (Child 1995) created value by converting raw materials into higher-value goods and services by replacing low-value meat production (i.e., beef or bushmeat) with high-value ecotourism. However, these gains rely on institutions, and when the rule of law breaks down (degrading institutions such as property rights, courts, arbitration, and economic exchange), people can no longer rely on impersonal institutions (North 2003). Personalized Big-Man economies re-exert themselves (Beinhocker 2006), and institutional breakdown often benefit elites (Chabal and Daloz 1999). The renewable resource economy, especially sectors like wildlife that rely on economies of scale and exchange, is quickly disrupted, and land use reverts to resources that people can own and protect individually like livestock or crops.

Similarly, communal lands are institutionally thin, lacking title deeds and genuine proprietary rights to wildlife and other natural resources like carbon. This locks in a reliance on privately owned livestock and individual (but untitled) fields even when these are low-value land use options. We can develop a rough estimate of the opportunity costs of weak institutions by comparing the output of communal and private drylands (400 to 600 mm annual rainfall) outside Kruger National Park, using data from Gregory Parent and Jessica Musengezi. In 2010, Parent conducted comprehensive surveys of production and consumption data from households in the Mutale area bordering northern Kruger National Park, in a study looking at the impacts of market access on household vulnerability and resource use decisions (unpublished data). His data show that conventional livestock and farming adds little or no value in terms of cash income or jobs (Table 7). The net output of South African Rand R232/ha from the local area is less than half of the requirements needed to maintain household consumption at R532/ha (Table 8). People are highly dependent on transfer payments from family members with jobs in urban areas and mines, and from government grants. Nonetheless, cropping, natural resource collection, and livestock are important risk mitigation and coping strategies, especially for poorer households. Households tend to view cattle more as an insurance policy than as a production commodity because they have limited access to formal insurance institutions.

Table 7: Annual output from land use activities in Benda Mutali communal area [i]

Activity	Total output	Output per hectareii	Output per household
Crops	97,185	4.19	171
Livestock	1,737,474	74.89	3,048
Natural resources	2,537,184	109.36	4,451
Labor	1,017,279	43.85	1,785
Total	5,389,122	232.29	9,455

[i]All values in South African Rand (US dollar $1 =R7).

[ii]Area based on mapping exercise with community tc calculate spatial extent of production, estimated to be 23,200 ha.

Table 8: Annual household consumption in Benda Mutali [i]

Activity	Total household consumption	Consumption per hectareii	Consumption per household
Crops	97,185	4.19	171
Livestock	768,152	33.11	1,348
Natural resources	1,364,392	58.81	2,394
Food purchases	5,242,040	225.95	9,197
Non-food purchases	4,207,784	181.37	7,382
Asset depreciation	718,040	30.95	1,260
Total	12,397,593	534.38	21,750

[i] All data in South African rand. (US dollar $1=R7).

[ii] Area based on mapping exercises with community to calculate spatial extent of production.

The key point is that, in the same agro-economic region, there is a 10 to 40 times difference in economic output between institutionally rich private land and institutionally thin communal land (noting also that communal lands suffer enormously from the legacies of apartheid). Communal lands also suffer disproportionately from overgrazing in a tragedy-of-the-commons-type situation where environmental costs are externalized to society (Hardin 1971). Our suggestion is that the opportunity costs of weak institutions are enormous, and the return on investment from developing new institutions may be high.

Policy 'Failure' in East Africa

The economic value of institutions is highlighted by the case of wildlife in East Africa. East Africa's wildlife is inherently superior to southern Africa, and the hunting sector only moved southwards when it was restricted in East Africa in the early 1970s. In contrast to southern Africa, postcolonial East Africa perpetuated colonial policies of state nationalisation, monopolisation, and centralization of wildlife (Kabiri 2010). Several private landholders such as David Hopcraft in Kenya have struggled to develop wildlife as a land use option for many years but suffer from weak proprietary rights, bureaucratic uncertainty, and restrictions on use. Thus, Hopcraft relies on tourism and the sale of meat to Nairobi for his income rather than on the potentially more lucrative safari hunting, which is banned in Kenya (personal communication). Some large landholders manage wildlife for tourism, but it would be a more competitive land use option in, for instance, the Laikipia district if hunting was legalized (Craig, personal communication). In Tanzania, hunting is legal, but the proceeds from hunting are seldom returned to landholder communities, while the state is also extracting most tourism revenues from community areas (Nelson 2010). Wildlife in prime areas in Kenya (e.g., Mara) has declined to a third or less of its former (1977) abundance (Ogutu et al. 2011), and there may well be similar downward trends in Tanzania, but data is sparse and often restricted.

Externalities, Collective Action, and Scale

Wildlife is a mobile resource that, like many natural resources, is associated with externalities, for example, where one landholder conserves a species only for his neighbour to hunt them or where wildlife damages crops on a neighbouring property. These externalities are the reason why individual ownership of wildlife is thought to be problematic and has been centralized in the state. However, externalities may be easier to solve in practice than theoretical approaches might suggest. This requires cross-scale institutions. Following the principle of subsidiarity, these need to be built up from the bottom. The primary goal is to internalize the costs and benefits of wildlife management at the level of landholders because they are deterministic of land use outcomes. The first essential step is to devolve proprietorship to landholders. Second, mechanisms are needed to manage externalities associated with the fugitive character of wildlife and other natural resources.

South Africa and Namibia manage wildlife's externalities with physical infrastructure. The right to utilize wildlife is linked to a 'certificate of adequate enclosure' and requires that wildlife properties are appropriately fenced. Game fencing solves problems associated with wildlife's mobility but introduces new problems because game fences are expensive and fragment landscapes. The sedentarisation of wildlife on small, fenced private ranches negates many natural behavioural advantages, allows overstocking, and results in the negative impacts on environments associated with decreased mobility in ruminants. Mismanagement of any animals, domestic or wild, causes ecological problems, and fencing savannas is a form of mismanagement. In the early days of game farming in South Africa, overstocking was a considerable concern, though the situation is improving with management experience and mechanisms for scaling up, such as conservancies.

The alternative to managing externalities with physical infrastructure is to manage them with institutions, as was done in Zimbabwe. Zimbabwe's far-sighted Natural Resources Act of 1941 enabled farmers to voluntarily establish themselves (usually 30 to 50 landholders in a catchment) as democratically-managed Intensive Conservation Areas (ICA) with more power to regulate themselves than anything a centralized state would dare to impose. ICAs had the power to require members to address soil erosion at their own financial cost, reduce livestock where overstocking was degrading the grass layer, desist from tree cutting, or set quotas and other restrictions on the use of wildlife species. Communities designed locally appropriate rules, which were enforced through peer pressure backed by the possibility of legal action. In essence, the community became the primary locus of regulation and enforcement. The state retained responsibility for regularly inspecting land to ensure that ICAs were performing and could intervene if this community-based system failed, but this was rarely required. The Act also established a court to which landholders could appeal.

ICAs became the focal point for social learning, collective action, and peer control of wildlife management, as farmers met to regulate and learn from each other. Self-regulation was parsimonious but effective. Landholders were financially and socially accountable for regulations they custom designed for their particular circumstances. They limited regulation to high-value species and circumstances with real benefits from regulation, and where regulations were socially acceptable and could be enforced. For example, in the Munyati-Sebakwe ICA, farmers set quotas for high-value species like sable and waterbuck that were still building up after the livestock era but did not waste their time regulating ubiquitous or territorial animals like impala and kudu. In the severe 1984 drought, the Chiredzi ICA acted collectively to save the few remaining sable in the area by capturing and feeding them (sable having been severely threatened by livestock overgrazing). As the species recovered, they judiciously increased hunting quotas, rewarding ranchers who successfully recovered their sable populations with higher quotas. In another example, one landholder was luring wildlife onto his land using fire and fodder to capture and sell them, but the ICA resolved this problem by imposing quotas on him. These issues were usually solved informally through peer pressure backed up by the ICA's power to impose collective solutions if they needed to. The system was inexpensive and highly effective. Decisions and enforcement were internally legitimated, managed primarily through peer pressure, and tightly designed to suit local circumstances.

The ICA system not only used social interaction to manage the externalities associated with natural resources but also provided a grassroots democratic system by which government officials were held accountable for natural resource policies. ICAs had direct statutory access to key ministers, parliament, and even the prime minister, and it is no coincidence that the movement was strongly supported by research, extension, and even environmental education both in schools and for adults (Child and Child, in preparation). Theoretically speaking, Zimbabwe's natural resource legislation combined privatization of wildlife to internalize costs and benefits at the level of the individual property, with collective action to control the spatial and temporal externalities associated with wildlife and natural resources.

Private Conservancies

In the 1990s, many ranchers began to recognize ecologies and economies of scale. They formed conservancies, whereby up to 30 ranchers agree, often contractually, to manage their wildlife collectively. This allowed a second economic transition. Whereas individual game ranchers could reintroduce rarer species like sable, hartebeest, and rhino, large wildlife collectives could restock big game like elephant, buffalo, lions, and even wild dogs. For example, the Save Conservancy in southeastern Zimbabwe captured over 600 elephants in the drought-stricken Gona-re-zhou National Park and restocked lions, wild dogs, buffalos, and plains game. In South Africa, landholder collectives removed the fence between themselves and Kruger National Park and generate substantial revenues from hunting and tourism. The economies of scale associated with tourism and wildlife allowed these huge private reserves to share management costs and principles over a larger area and to generate more money than individual properties.

Community-Based Natural Resource Management (CBNRM)

From the early 1970s, conservation bureaucrats in Zimbabwe, Namibia, and Botswana (and also Zambia) argued that it was unfair that only white farmers should benefit from wildlife and that, if similar rights were devolved to communal areas, both rural people and wildlife would benefit (Metcalfe 1993). This led to the emergence of community-based natural resource management (CBNRM) in southern Africa where, led by Zimbabwe's iconic CAMPFIRE programme (Child 1993;Murphree 2005; Frost and Bond 2008; Taylor 2009), hunting and tourism revenues were returned to communities. From a conservation perspective, this stabilised or even increased wildlife populations in Zimbabwe's crowded communal lands (Child et al. 2003), and the recovery of wildlife, including endangered rhinos and elephants, was even stronger in Namibia (NACSO 2008). Income from wildlife contributed to social projects, provided cash to households in some projects (Dalal-Clayton and Child 2003; Jones and Weaver 2009), and helped to empower local people (Jones and Murphree 2004). In some communities, employment and income from hunting and tourism contributed significantly to livelihoods, providing as much as three-quarters of on-farm livelihoods to the Sankuyo community in Botswana (unpublished data).

Community-based natural resource management (CBNRM) is the subject of a considerable literature, and we emphasize only key issues here. First, wildlife-based CBNRM is founded on the potential to use wildlife as an economic engine in communal lands and, in so doing, to incentivise conservation. Second, CBNRM mostly does not involve a trade-off between conservation and development because it is an institutional approach that allocates natural resources including wildlife to higher valued uses, and in some cases, more equitably. Thus, CBNRM can and should simultaneously contribute to both conservation and development, as happened for private land. Third, CBNRM relies on two sets of institutions: those that devolve rights (or not) from state and community and about which much is written (Murphree 2004; Nelson 2010), and those within a community that promote participation and equitable benefit sharing, which is an emerging challenge that has hardly been studied.

As with private conservation, an essential ingredient of CBNRM is the devolution of the rights and benefits for wildlife to communities. Where all (e.g., Botswana and Namibia) or most (e.g., Zimbabwe) income has been devolved to communities, CBNRM has encouraged the recovery of wildlife and generated significant incomes (NACSO 2008; Taylor 2009), but has not magically (and unrealistically) resolved poverty, because of the number of people involved and because wildlife populations are still low and recovering in hot, dry, and remote environments. CBNRM is also difficult to implement when human densities are high, wildlife populations are depleted, and benefits are low. However, the so-called 'crisis' in CBNRM reflects a failure of implementation and, specifically, a failure to devolve rights and responsibilities to communities within a carefully designed institutional framework. Many, if not most CBNRM programmes, do not represent a serious attempt to devolve revenues to communities and therefore cannot be expected to work (Nelson 2010; Ribot 2003). These are CBNRM programmes in name only, and wildlife is neither benefiting communities nor recovering therefore.

The emerging challenge for CBNRM, and about which little is written, is the challenge of within-community governance. Commonly CBNRM programmes generate revenue and recover wildlife populations, but like many decentralization projects, they are plagued by elite capture, misuse of money, and low levels of participation. In our experience, this is because they rely on representational forms of governance whereby leaders elected by a community sit on committees that make many or most of the decisions on behalf of the community. The assumption is flawed that bi- or tri-annual elections and annual general meetings will achieve accountability, participation, transparency, and equitable benefit sharing (and guard against financial misappropriation and elite capture).

Challenging as it may seem, participatory democracy may be an essential starting point and building-block for a successful CBNRM (and much easier to implement than first assumed). Real participation requires that CBNRM communities are defined at a scale and implement participatory processes that allow the majority of the people to (1) meet face-to-face to share information and make decisions, (2) instruct the committee to implement these decisions, and (3) review the quality of implementation (i.e., does expenditure reflect the budget agreed upon by the community?) no less than quarterly.

Scaling Up and Scaling Down at the Same Time

The experience of private game ranching demonstrates that wildlife can greatly increase the economic output from drylands. We can have confidence in wildlife utilization as an economic model. However, we need to develop new governance models to substantially increase levels of participation and benefit sharing, and that prevent money from being diverted to administrative expenses and elite capture. We suggest that this requires CBNRM programs to be scaled down to the level of single-village communities and processes of participatory democracy be strictly followed. In other words, CBNRM requires carefully crafted institutions to manage the challenge of scale. These institutions need to scale up for reasons of ecological and economic scale, but also to simultaneously scale down to improve accountability, participation, and equitable benefit sharing (Child, unpublished data).

Hunting

One of the more controversial and misunderstood aspects of wildlife utilization is hunting. Trophy hunting is an ecologically robust and high-value use of wildlife. It is true that non-hunting tourism can out compete hunting tourism economically but only in prime areas with excellent wildlife. Hunting remains the bread and butter of wildlife operations, especially in non-prime areas and in times of global or national crisis when tourists stop traveling but hunters do not. Hunting provides a significant cash flow from very few animals, facilitating the transition towards wildlife-based land uses. Indeed, it is difficult to make the transition from livestock to wildlife in the absence of hunting, except for people with large amounts of money to invest. We see this even in the development of Pilanesberg Game Reserve in South Africa (Johnson et al. 2009). The government purchased degraded livestock ranches to form a new park on the basis that this was a better economic use of the land. The transition was partly financed by carefully controlled trophy hunting of excess males. Hunting was carefully zoned in time and place to avoid conflicts with the rapidly growing photo-safari business while careful hunting practices were used, so wildlife did not associate gunshots with vehicles or people.

Hunting is an essential component of the transition to wildlife-based land uses. However, as wildlife populations and diversity recover, landholders are able to introduce photographic tourism, building attractive lodges, and training significant numbers of local guides, cooks, caterers, and managers. However, even in these circumstances, photo-tourism and hunting are compatible. Conflicts are certainly common between tour operators and hunting outfitters when they are competing for land and wildlife (e.g., around the Okavango Delta in Botswana and the Luangwa Valley in Zambia), but such problems seldom arise when the same manager is responsible for both operations, as in the Pilanesberg example.

CONCLUSIONS

It has taken 60 years for wildlife to emerge as an important and economically viable land use option. It is now a proven economic model, and given the right ecological and institution circumstance, wildlife can provide significantly more jobs and economic growth than conventional uses of the same land. We still face the challenge of how to apply this model more broadly in Africa's communal lands, a challenge that requires genuine devolution of proprietorship from government to community and democratic and effective organizational development within communities.

The wildlife option has emerged through proactive policy making and private sector entrepreneurship. Far-sighted conservation bureaucrats challenged the postcolonial status quo and developed a new approach to wildlife conservation based on landholder benefit. They sought to maximize the value from wildlife to people living with it by (1) devolving full proprietorship to landholders, (2) encouraging sustainable commercial uses provided they were humane, and (3) reducing the regulatory burden on the sector. This allowed thousands of wildlife owners to experiment with new products, steadily driving up the value of wildlife in ways that we seldom see with other wild resources like forests. Where institutions were changed in this way, mainly on private land but also in some communal lands, wildlife and the habitats on which it depends have recovered enormously. But where wildlife remained centralized and noncommercial, wildlife populations have declined.

Many drylands are locked in a reinforcing cycle of ecological over-utilization, but economic under-utilization often linked to a reliance on commodity-based production systems. Bold policy facilitated a transformation to an economy based not on commodities but on experiences - the wildlife and bio-experience sector. This generates more economic activity yet extracts far less from the environment and, though it seems counterintuitive in a world afflicted with poverty, the terms of trade for tourism and experiences are improving steadily whereas the price of agricultural commodities has been stagnant or falling for several decades.

In the past few years, recognition of the importance of ecosystem services has increased. In the most thorough analysis to date, Robert Constanza and his colleagues demonstrated that the value of ecosystems exceeds global GDP by several factors (Costanza, d'Arge et al. 1997). But the problem is that these values are not reflected in the private prices that guide the decision makers of landholders nor, therefore, in land use outcomes. In the 1960s, wildlife faced these very same problems. They were addressed by adopting a new approach to conservation based on well-crafted institutions at both the national and local level that sought, first, to maximize the value of wildlife to the landholder by internalizing (privatizing) its benefits and costs at the level of the landholder, and to simultaneously establish democratic collective action to manage externalities and scale. Contrary to the predictions of conservationists, commercializing wildlife in this way has had positive conservation and development outcomes. Positive outcomes hinge on the simultaneous application of two factors: driving up the price of wildlife (i.e., price) while ensuring that benefits and management are controlled by landholders (i.e., proprietorship and subsidiarity). In the absence of proprietorship, for example, the high price of ivory and rhino horn has been disastrous, but taking away the value of wildlife is not a solution either because it is quickly replaced by alternative land uses.

The development of the wildlife sector has been a policy experiment based on the hypothesis that giving wildlife value to the people who live with it will result in positive development and conservation outcomes, requiring considerable tenacity and adaptive learning in navigating towards this objective in complex economic and political circumstances. The approach has succeeded where laws, regulations, and operational procedures have incorporated the principles of price, proprietorship, subsidiarity, and adaptive management in a theoretically sound and disciplined manner. Indeed, wildlife provides a pioneering example of Payments for Ecosystem Services, which was relatively easy to implement because of the high value of wildlife and the relative simplicity of costs and benefits attribution (at least compared to ecosystem processes). Ecosystem services are facing similar market failures of wildlife. The challenge of our generation may well be to convert these values into incentives for maintaining ecosystem services, perhaps by adapting some of the lessons from the wildlife sector in southern Africa.

AUTHORS' CONTRIBUTIONS

BC conceptualized and compiled the paper. JM provided data on game ranching and GP provided data on community livelihoods in South Africa. GC strengthened the paper's historical and policy perspectives. All authors read and approved the final manuscript.

REFERENCES

1. Astle, WL. 1999. A History of wildlife conservation and management in the mid-Luangwa Valley, Zambia. British Empire and Commonwealth Museum, Bristol.
2. Barnes, J, and B Jones. 2009. Game ranching in Namibia. In Evolution and innovation in wildlife conservation, ed. Suich H, Child B, 113–126. Earthscan, London.
3. Beinhocker, ED. 2006. The origin of wealth: evolution, complexity and the radical remaking of economics. Harvard Business School Press, Boston.
4. Bell, RHV. 1982. The effect of soil nutrient availability on community structure in Africa savannas. In Ecology of tropical savannas, ed. Huntley BJ, Walker BH, 193–216. Springer verlag, Berlin.
5. Booth, V. 2002. Analysis of wildlife markets (sport hunting and tourism). WWF Southern African Regional Programme Office, Harare.
6. Booth, VR. 2009. A comparison of the prices of hunting tourism in southern and eastern Africa. Joint publication of Food and Agriculture Organization of the United Nations (FAO) and International Council for Game and Wildlife Conservation (CIC), Budapest.
7. Borgerhoff Mulder, M, and P Coppolillo. 2005. Conservation: linking ecology, economics, and culture. Princeton University Press, Princeton.
8. Bothma, , Jd P, H Suich, and A Spenceley. 2009. Extensive wildlife production on private land in South Africa. In Evolution & innovation in wildlife conservation, ed. Suich H, Child B, 147–162. Earthscan, London.
9. Brockington, D, and J Igoe. 2006. Eviction for conservation: a global overview. Conservation and Society 4 (3): 424–470.
10. Campbell, A, and G Child. 1971. The impact of man on the environment of Botswana. Botswana Notes and Records 3: 91–110.
11. Carruthers, J. 1989. Creating a national park 1910–1926. Journal of Southern African Studies 15: 188–216.

12. Carruthers, J. 2008. "Wilding the farm or farming the wild"? The evolution of scientific game ranching in South Africa from the 1960s to the present. Transactions of the Royal Society of South Africa 63 (2): 160–181.
13. Chabal, P, and JP Daloz. 1999. Africa works: the political instrumentalization of disorder. James Currey, Oxford.
14. Chadwick, DH. 1996. A place for parks in the new South Africa. National Geographic 190 (1): 2–41.
15. Child, G. 1971. The future of wildlife and rural land use in Botswana. Proceedings of the SARCCUS Symposium. Nature Conservation as a Form of Land Use, Gorongoza National Park, Mozambique.
16. Child, B. 1988. The role of wildlife utilization in the sustainable economic development of semi-arid rangelands in Zimbabwe. PhD thesis. University of Oxford.
17. Child, B. 1993. Zimbabwe's CAMPFIRE programme: using the high value of wildlife recreation to revolutionize natural resource management in communal areas. Commonwealth Forestry Review 72 (4): 284–296.
18. Child, G. 1995. Wildlife and people: the Zimbabwean success. How the conflict between animals and people became progress for both. Wisdom Foundation, Harare.
19. Child, B. 2004. The Luangwa integrated rural development project, Zambia. In Rights, resources and rural development. Community-based natural resource management in Southern Africa, ed. Fabricius C, Kock E, Magome H, Turner S, 235–247. Earthscan, London.
20. Child, B. 2004b. ed. Child B, Lyman MW, 17–50. Sand County Foundation and Washington: The Aspen Institute, Madison.
21. Child, B. 2006. Revenue distribution for empowerment and democratization. In Participatory Learning and Action 55: 20–29.
22. Child, B. 2009. Game ranching in Zimbabwe. In Evolution and innovation in wildlife conservation, ed. Suich H, Child B, 127–146. Earthscan, London.
23. Child, B, and G Barnes. 2010. The conceptual evolution and practice of CBNRM in southern Africa – past, present and future. Environmental Conservation 37 (3): 83–295.

24. Child, GFT, and T Riney. 1987. Tsetse control hunting in Zimbabwe, 1919–1958. Zambezia 14 (1): 11–72.

25. Child, B, B Jones, D Mazambani, A Mlalazi, and H Moinuddin. 2003. Final evaluation report: Zimbabwe natural resources management program - USAID/Zimbabwe Strategic Objective No. 1. CAMPFIRE Communal Areas Management Programme for Indigenous Resources, 153. USAID, Washington.

26. Coe, MJ, DH Cumming, and J Phillipson. 1976. Biomass and production of large African herbivores in relation to rainfall and primary production. Oecologia 22: 341–354.

27. Costanza, R, d' Arge R, R de Groot, S Farberk, M Grasso, H Bruce, K Limburg, S Naeem, RV O'Neill, J Paruelo, RG Raskin, S Paul, and M vanden Belt. 1997. The value of the world's ecosystem services and natural capital. Ecological Economics 25: 3–15.

28. Cousins, JA, JP Sadler, and J Evans. 2008. Exploring the role of private wildlife ranching as a conservation tool in South Africa: stakeholder perspectives. Ecology and Society 3 (2): 1 34.http://www.ecologyandsociety.org/vol13/iss2/art43/

29. Cumming, DHM. 1982. The influence of large herbivores on savanna structure in Africa. In Ecology of tropical savannas, ed. Huntley BJ, Walker BH, 217–245. Springer, New York.

30. Cumming, D. 1995. Are multispecies systems a viable land use option for southern African rangelands? In Wild and domestic ruminants in extensive land use systems, ed. Hofmann RR, Schartz HJ, 203–234. Humboldt University, Berlin.

31. Cumming, DHM, and I Bond. 1991. Animal production in southern Africa: present practices and opportunities for peasant farmers in arid lands. Multispecies animal production systems project paper no. 22. WWF Multispecies Project, Harare.

32. Dalal-Clayton, B, and B Child. 2003. Lessons from Luangwa. The story of the Luangwa Integrated Resource Development Project, Zambia. International Institute for Environment and Development, London.

33. Dasmann, RF. 1964. African game ranching. Pergamon Press, Oxford.

34. Dasmann, RF, and AS Mossman. 1961. Commercial utilization of game animals on a Rhodesian ranch. National Museums, Salisbury.

35. Dry, G. 2010a. Why game farming should be taken seriously. Farmer's Weekly, Pretoria, South Africa.14 May 2010
36. Dry, GC. 2010. Commercial wildlife ranching's contribution to the "green economy". Green Economy Summit, South Africa.
37. Frost, PGH, and I Bond. 2008. The CAMPFIRE programme in Zimbabwe: payments for wildlife services. Ecological Economics 65 (1): 776–787.
38. Grossman, D, and M Gandar. 1989. Land transformation in South African savanna regions. South African Geographical Journal 7 (1): 38–45.
39. Hardin, GJ. 1971. The tragedy of the commons. Science 162: 1243–1248.
40. Heijnsbergen, VP. 1997. International legal protection of wild fauna and flora. Landsdale IOS Press, Amsterdam.
41. Hulme, D, and M Murphree. 2001. Community conservation in Africa. An introduction. In African wildlife and livelihoods The promise and performance of community conservation, ed. Hulme D, Murphree M, 1–37. James Currey, Oxford.
42. IUCN. 1963. Conservation of nature and natural resources in modern African states. IUCN Publications new series. International Union for the Conservation of Nature and Natural Resources, Morges.
43. Jansen, DJ, I Bond, and B Child. 1992. Cattle, wildlife, both or neither? A survey of commercial ranches in the semi-arid regions of Zimbabwe. WWF Multispecies Animal Production Project, 203 plus appendices, Harare.
44. Jarman, PJ. 1973. The social organization of antelope in relation to their ecology. Behaviour 48 (3/4): 215–267.
45. Jenkins, H. 2011. Opening address, 7th International Wildlife Ranching Symposium by Premier of the Northern Cape. 7th International Wildlife Ranching Symposium, Kimberly, South Africa.
46. Johnson, SR, W Boonzaaier, R Collinson, and R Davies. 2009. Changing institutions to respond to challenges: North West Parks, South Africa. In Evolution and innovation in wildlife conservation, ed. Suich H, Child B, 289–306. Earthscan, London.

47. Jones, BTB, and MW Murphree. 2004. Community-based natural resources management as a conservation mechanisam: lessons and directions. Parks in transition: biodiversity, rural development and the bottom line. ed. Child B, 63–104. Earthscan, London.

48. Jones, B, and C Weaver. 2009. CBNRM in Namibia: growth, trends, lessons and constraints. In Evolution & innovation in wildlife conservation, ed. Suich H, Child B, 223–242. Earthscan, London.

49. Kabiri, N. 2010. Historical and contemporary struggles for a local wildlife governance regime in Kenya. In Community rights, conservation & contested land. The politics of natural resource governance in Africa, ed. Nelson F, 121–144. Earthscan, London.

50. Kelly, RD, and BH Walker. 1976. The effects of different forms of land use on the ecology of a semi-arid region in South-Eastern Rhodesia. Journal of Ecology 64 (2): 553–576.

51. Langholz, JA, and GIH Kerley. 2006. Combining conservation and development on private lands: an assessment of ecotourism-based private game reserves in the Eastern Cape. Nelson Mandela Metropolitan University, Center for African Conservation Ecology.31

52. Libecap, GD. 2009. The tragedy of the commons: property rights and markets as solutions to resource and environmental problems. Australian Journal of Agricultural and Resource Economics 53 (1): 129–144.

53. Lindsay, P, R Du Toit, A Pole, and S Romañach. 2009. Save Valley conservancy: a large scale African experiment in cooperative wildlife management. In Evolution & innovation in wildlife conservation, ed. Suich H, Child B, 163–186. Earthscan, London.

54. Lindsey, PA, GA Balme, et al.. 2012. The Significance of African Lions for the Financial Viability of Trophy Hunting and the Maintenance of Wild Land. PLoS Biol 7 (1):

55. Mahesh Sankaran, NPHRJS, J Ratnam, DJ Augustine, BS Cade, J Gignoux, SI Higgins, X Le Roux, F Ludwig, J Ardo, F Banyikwa, A Bronn, G Bucini, KK Caylor, MB Coughenour, A Diouf, W Ekaya, CJ Feral, EC February, PGH Frost, P Hiernaux, H Hrabar, KL Metzger, HHT Prins, S Ringrose, W Sea, J Tews, J Worden,

and N Zambatis. 2005. Determinants of woody cover in African savannas. Nature 438: 846–849.

56. Martin, R. 1986. Communal areas management programme for indigenous resources. Revised edition. Department of Natural Parks and Wildlife Management, Branch of Terrestrial Ecology, Harare.

57. Martin, R. 2009. From sustainable use to sustainable development. Evolving concepts of natural resource management. IUCN - Southern African Sustainable Use Specialist Group 55.

58. Metcalfe, S. 1993. CAMPFIRE - Zimbabwe's communal areas management programme for indigenous resources. In Natural connections: perspectives in community-based conservation, ed. Western D, Wright M, Strum S, Island Press, Washington.

59. Monke, EA, and SR Pearson. 1989. The policy analysis matrix for agricultural development. Cornell University Press, Ithaca.

60. Mossman, AS. 1975. International game ranching programs. Journal of Animal Science 40: 993–999.

61. Muir-Leresche, K, and RH Nelson. 2000. Private property rights to wildlife: the southern African experiment. International Centre for Economic Research, Italy.

62. Murphree, M. 1994. Communities as resource management institutions. International Institute for Environment and Development Gatekeeper Series, London.No 36

63. Murphree, MW. 2004. Communal approaches to natural resource management in Africa: from whence to where? In Breslauer Symposium on Natural resource Issues in Africa. University of California, Berkeley.

64. Murphree, M. 2005. Congruent objectives, competing interests, and strategic compromise: concept and process in the evolution of Zimbabwe's CAMPFIRE, 1984–1996. Communities and conservation: histories and politics of community-based natural resource management. ed. Brosius JP, Tsing AL, Zerner C, 105–148. Rowman and Littlefield, Oxford.

65. Musengezi, J. 2010. Wildlife utilization on private land: understanding the economics of game ranching in South Africa. PhD thesis. University of Florida.

66. NACSO. 2006. Namibia's communal conservancies. A review of progress and challenges in 2005. Namibian Association of CBNRM Support Organisations 104, Windhoek.
67. NACSO. 2008. Namibia's communal conservancies. A review of progress and challenges in 2007. Namibian Association of CBNRM Support Organizations 120, Windhoek.
68. Nelson, F. 2010. Introduction: the politics of natural resource governance in Africa. In Community rights, conservation and contested land: the politics of natural resource governance in Africa, ed. Nelson F, 3–31. Earthscan, London.
69. Nelson, F, and T Blomley. 2009. Peasants' forests and the king's game? Institutional divergence and convergence in Tanzania's forestry and wildlife sectors. In Community rights, conservation and contested land: the politics of natural resource governance in Africa, ed. Nelson F, Earthscan, London.
70. North, DC. 1990. Institutions, institutional change and economic performance. Cambridge University Press, Cambridge.
71. North, DC. 1995. The new institutional economics and third world development. In The new institutional economics and third world development. In , ed. Harriss J, Hunter J, Lewis CM, 17–26. Routledge, London.
72. North, DC. 2003. Understanding the process of economic change. In Forum series on the role of institutions in promoting economic growth: June 24, 2003. Mercatus Center at George Mason University and The IRIS Center, Washington.
73. Ogutu, JO, N Owen-Smith, et al.. 2011. Continuing wildlife population declines and range contraction in the Mara region of Kenya during 1977–2009. Journal of Zoology 283 (2): 99–109.
74. Parker, I. 2004. What I tell you three times is true: conservation, ivory, history and politics. Librario, UK.
75. Phimister, IR. 1978. Meat and monopolies: beef cattle in southern Rhodesia, 1890–1938. The Journal of African History 19: 391–414.
76. PriceWaterhouse. 1994. The Lowveld conservancies: new opportunities for production and sustainable land-use. Save, Bubiana and Chiridzi River Conservancies, Harare.

77. Ribot, JC. 2003. Democratic decentralisation of natural resources: institutional choice and discretionary power transfers in sub-Saharan Africa. Public Administration and Development 23 (1): 53–65.
78. Ribot, JC, JF Lund, and T Treue. 2010. Democratic decentralization in sub-Saharan Africa: its contribution to forest management, livelihoods, and enfranchisement. Environmental Conservation 37 (1): 35–44.
79. Riney, T. 1960. Rhodesian wildlife is a natural resource. Wild Life 2 (1): 49–54.
80. Riney, T. 1963. A rapid field technique and its application in describing conservation status and trends in semi-arid pastoral areas. African Soils 8: 159–257.
81. Riney, T, and P Hill. 1967. Conservation and management of African wildlife. Food and Agriculture Organization FAO, Rome.
82. SASUSG. 1996. Sustainable use issues and principles. Southern Africa Sustainable Use Specialist Group, IUCN Species Survival Commission 23.
83. 2009. Evolution & innovation in wildlife conservation. Parks and game ranches to transfrontier conservation areas,Suich H, Child B Earthscan, London.
84. Talbot, LM, HP Ledger, and WJA Payne. 1961. The possibility of using wild animals for animal production on East African rangelands based on a comparison of ecological requirements and efficiency of range utilization by domestic livestock and wild animals. In Reports of the 8th International Congress of Animal Production. Eugen Elmer Verlag, Hamburg.
85. Talbot, LM, WJA Payne, HP Ledger, LD Verdcourt, and MH Talbot. 1965. The meat production potential of wild animals in Africa. A review of the biological knowledge. Technical Communication No. 16, Commonwealth Bureau Animal Breeding and Genetics vol 16. Commonwealth Agricultural Bureaux, Edinburgh.
86. Taylor, S. 2002. Save Valley conservancy [film]. Zimbabwe.60 minutes
87. Taylor, R. 2009. Community based natural resource management in Zimbabwe: the experience of CAMPFIRE. Biodiversity and Conservation 18: 2563–2583.

88. Taylor, RD, and RB Martin. 1987. Effects of veterinary fences in Zimbabwe. Environmental Management 11 (3): 327–334.

89. Taylor, RD, and BH Walker. 1978. Comparison of vegetation use and herbivore biomass on a Rhodesian game and cattle ranch. Journal of Applied Ecology 15: 565–581.

90. Vincent, V, and RG Thomas. 1961. An agricultural survey of southern Rhodesia. Federation of Rhodesia and Nyasaland, Salisbury.2 Volumes

91. 1987. Determinants of tropical savannas,Walker BH IRL, Oxford.

92. Walker, B, CS Holling, SR Carpenter, and A Kinzig. 2004. Resilience, adaptability and transformability in social-ecological systems. Ecology and Society 9 (2): 2–10.

93. Williams, TO. 1993. Livestock pricing policy in sub-Saharan Africa: objectives, instruments and impact in five countries. Agriculture Economics 8: 139–159.

Chapter 3

Risks And Precautions of Genetically Modified Organisms

Dhan Prakash[1], Sonika Verma[2], Ranjana Bhatia[1], and B. N. Tiwary[3]

[1]Institute of Microbial Technology (CSIR), Sector 39A, Chandigarh 160036, India

[2]Department of Biotechnology, UIET, Punjab University, Chandigarh, India

[3]Department of Biotechnology, Guru Ghasidas Vishwavidyalaya (A Central University), Bilaspur 495009, India

ABSTRACT

Commercial potential of biotechnology is immense since the scope of its activity covers the entire spectrum of human life. The most potent biotechnological approach is the transfer of specifically constructed gene assemblies through various techniques. However, this deliberate modification and the resulting entities thereof have become the bone of contention all over the world. Benefits aside, genetically modified organisms (GMOs) have always been considered a threat to environment and human health. In view of this, it has been considered necessary by biosafety regulations of individual countries to test the feasibility of GMOs in contained and controlled environments for any potential risks they may pose. This paper describes the various aspects of risk, its assessment, and management which are imperative in decision making regarding the safe use of GMOs. Efficient efforts are necessary for implementation of regulations. Importance of the risk assessment, management, and precautionary approach in environmental agreements and activism is also discussed.

INTRODUCTION

Modern biotechnology has allowed the movement of genetic material across unrelated species, something impossible with the traditional breeding methods. This intentional transfer of genetic material has in turn brought biotechnology out from the laboratory to the field. Genetically modified organisms (GMOs) are organisms whose genetic material has been artificially modified to change their characteristics in some way or another [1]. In essence, "genetic modification" or "genetic engineering" techniques enable scientists to find individual genes that control particular characteristics, separate them from the original source, and transfer them directly into the cells of an animal, plant, bacterium, or virus. This technology has many potential applications [2]. These new opportunities bring along greater public scrutiny and government regulation. Risk assessment is a common regulatory tool used in the decision-making process for a proposed commercial release of a GMO into the environment [3, 4].

Environmental applications of microorganisms are wide and varied, ranging from bioremediation, biopesticides, nitrogen fixation, plant growth promoter, to biocontrol of plant diseases, and other such agricultural practices. The sensible application of recombinant DNA techniques has shown the potential for genetically improved microorganisms to be used as soil or seed inoculants [5–8]. However, when introduced into the environment, they could have unintended environmental consequences and may play more pronounced ecological roles than the wild types [9–11]. Genetically improved microorganisms are able to reproduce and establish themselves as persistent populations and may have subtle and long-term effects on biological communities and natural ecosystems [12]. Results of DNA modification may not be limited only to the particular characteristics of the replaced gene. It is therefore important to ensure that when these organisms are released into nature they do not harm the environment or human health [13]. Such concerns have led to broader interests in the theme of risk assessment in the release of GMOs. A cautious approach is necessary to assess environmental risks which may occur due to introduction of recombinant organisms in the natural environment [14].

RISKS RELATED TO THE USE OF GENETICALLY MODIFIED ORGANISMS ECOLOGICAL STABILITY OF THE GMO

The application of genetic modification allows genetic material to be transferred from any species into plants or other organisms. The introduction of a gene into different cells can result in different outcomes, and the overall pattern of gene expression can be altered by the introduction of a single gene. The sequence of the gene and its role in the donor organism may have a relatively well-characterized function in the organism from which it is isolated. However, this apparent "precision" in the understanding of a gene does not mean that the consequences of the transfer are known or can be predicted [15]. Copies of a gene may be integrated, additional fragments inserted, and gene sequences rearranged and deleted—which may result in lack of operation of the genes instability or interference with other gene functions possibly cause some potential risks [16]. Therefore, there could be a number of predictable and unpredictable risks related to release of GMOs in the open environment. The report prepared by the Law Centre of IUCN, the World Conservation Union (2004), enlists numerous environmental risks likely to occur by the use of GMOs in the field. These risks are as follows.

Each gene may control several different traits in a single organism. Even the insertion of a single gene can impact the entire genome of the host resulting in unintended side effects, all of which may not be recognizable at the same time. It is difficult to predict this type of risk.

Genetic Contamination/Interbreeding

Introduced GMOs may interbreed with the wild-type or sexually compatible relatives. The novel trait may disappear in wild types unless it confers a selective advantage to the recipient. However, tolerance abilities of wild types may also develop, thus altering the native species' ecological relationship and behaviour.

Competition with Natural Species

Faster growth of GMOs can enable them to have a competitive advantage over the native organisms. This may allow them to become invasive, to spread into new habitats, and cause ecological and economic damage.

Increased Selection Pressure on Target and Nontarget Organisms

Pressure may increase on target and nontarget species to adapt to the introduced changes as if to a geological change or a natural selection pressure causing them to evolve distinct resistant populations.

Ecosystem Impact

The effects of changes in a single species may extend well beyond to the ecosystem. Single impacts are always joined by the risk of ecosystem damage and destruction.

Impossibility of Followup

Once the GMOs have been introduced into the environment and some problems arise, it is impossible to eliminate them. Many of these risks are identical to those incurred with regards to the introduction of naturally or conventionally bred species. But still this does not suggest that GMOs are safe or beneficial, nor that they should be less scrutinized.

Horizontal Transfer of Recombinant Genes to Other Microorganisms

One risk of particular concern relating to GMOs is the risk of horizontal gene transfer (HGT). HGT is the acquisition of foreign genes (via transformation, transduction, and conjugation) by organisms in a variety of environmental situations. It occurs especially in response to changing environments and provides organisms, especially prokaryotes, with access to genes other than those that can be inherited [17, 18].

HGT of an introduced gene from a GMO may confer a novel trait in another organism, which could be a source of potential harm to the health of people or the environment. For example, the transfer of antibiotic resistance genes to a pathogen has the potential to compromise human or animal therapy [19]. HGT has been observed for many different bacteria, for many genes, and in many different environments. It would therefore be a mistake to suppose that recombinant genes would not spread to other bacteria, unless precautions are taken. Recent evidence from the HGT technology confirms that transgenic DNA in GM crops and products can spread by being taken up directly by viruses and bacteria as well as plant and animals cells. Very recently, Yoshida et al. [20] reported that HGT also moved from a nuclear monocot gene into the genome of the eudicot parasite witchweed, which infects many grass species in Africa.

Some of the important potential impacts of HGT from GMOs include the following [21].

Adverse Effects on the Health of People or the Environment

These include enhanced pathogenicity, emergence of a new disease, pest or weed, increased disease burden if the recipient organism is a pathogenic microorganism or virus, increased weed or pest burden if the recipient organism is a plant or invertebrate, and adverse effects on species, communities, or ecosystems.

Unpredictable and Unintended Effects

HGT may transfer the introduced genes from a GMO to potential pests or pathogens and many yet to be identified organisms. This may alter the ecological niche or ecological potential of the recipient organism [9] and even bring about unexpected changes in structure or function [22]. Furthermore, the gene transferred may insert at variable sites of the recipient gene, not only introducing a novel gene but also disrupting an endogenous gene, causing unpredictable and unintended effects.

Loss of Management Control Measures

Regulatory approvals for field trials of GMOs often require measures to limit and control the release in space and time. With the spread of the introduced gene(s) to another species by HGT, a new GMO is created. This new GMO may give rise to adverse effects which are not controlled by management measures imposed by the original license or permit.

Long-Term Effects

Sometimes the impact of HGT may be more severe in the long term. Even under relatively strong selection pressure, it may take thousands of generations for a recipient organism to become the dominant form in the population [23]. In addition, other factors such as timing of appropriate biotic or abiotic environmental conditions and additional changes in the recipient organism could delay adverse effects.

Ethical Concerns

Various ethical issues associated with HGT from GMOs have been raised including perceived threats to the integrity and intrinsic value of the organisms involved, to the concept of natural order and integrity of species, and to the integrity of the ecosystems in which the genetically modified organism occurs [24].

Several scientific evidence that has emerged on GMOs over the last couple of years shows that there are several clear risks to human health and the environment. When genetic engineers create GMO or transgenic plants, they have no means of inserting the gene in a particular position. The gene ends up in a random location in the genetic material, and its position is not usually identified [25, 26]. There are already several examples of such undesired effects being identified in the US after approval (e.g., GM cotton with deformed cotton bolls; increased lignin in GM soya, etc.) [27]. Releasing genetically modified plants or crop into the environment may have direct effects, including gene transfer to wild relatives or conventional crops, weediness, trait effects on nontarget species, and other unintended effects [28].

It is widely accepted that the gene flow from GM crops is possible through pollen, from open-pollinated varieties crossing with local crops or wild relatives [29]. Because gene flow has happened for millennia between land races and conventionally bred crops, it is reasonable to expect that it could also happen with transgenic crops. Transgenic crops vary in their tendency to outcross, and the ability to outcross depends on the presence of sexually compatible wild relatives or crops, which varies according to location. However, some lines of evidence suggested that whether or not gene flow between transgenic crops and wild relatives matters, in and of itself [15]. If a resulting transgenic/wild hybrid had some competitive advantage over the wild population, it could persist in the environment and potentially disrupt the ecosystem [28, 30].

In addition, some indirect effects of GMO were also observed which potentially harm to the environment. For example, some transgenic traits such as the pesticidal toxins expressed by Bt genes may affect nontarget species as well as the crop pests. It could happen but still uncertain how likely it is [31, 32]. The toxicological studies of Monarch butterfly provide excellent examples, which established the sensitivity of Monarch larvae to consuming Cry1Ab protein from Bacillus thuringiensis (Bt) expressed in transgenic maize [33], thereby triggering further to assess exposure and population level effects [31]. It was determined that larval exposure to pollen on a population-wide basis was low, given the proportion of larvae in maize fields during pollen shed, the proportion of fields planted in Bt maize, and the levels of pollen within and around maize fields that exceed the toxicity threshold [29, 30]. However, an acute dose, even if several times higher than would be expected in the field, is not equivalent to a low natural chronic dose experienced over a longer period; therefore, a two-year study was undertaken and subsequently demonstrated that the risk to Monarch butterfly populations is 0.6% of the total of Monarch butterflies breeding in the North American Corn Belt [31, 34]. These results indicated negligible effects of Bt pollen to Monarch butterfly larvae from extended exposures in field settings.

Extensive long-term use of herbicides glyphosate and gluphosinate in the Bt crops can promote the development of resistant insect pests and weeds. The Royal Society in the year 2003 has published the results of extensive farm-scale evaluations of the impacts of transgenic HT maize, spring oilseed rape (canola), and sugar beet on biodiversity in the United Kingdom. These studies found that the main effect of these crops compared with conventional cropping practices was on weed vegetation, with consequent effects on the herbivores, pollinators, and other populations that are feed on it. These groups were negatively affected in the case of transgenic HT sugar beet, were, positive in case of HT Maiza and showed no effect in spring oilseed rape. However, there is still insufficient evidence to predict what the long-term impacts of transgenic HT crops will be on weed populations and associated in-crop biodiversity.

Most of the ecologists agree that gene flow is not an environmental problem unless it leads to undesirable consequences. In the short term, the spread of transgenic herbicide resistance via gene flow may create logistical and/or economic problems. Over the long term, transgenes that confirm resistance to pests and environmental stress and/or lead to greater seed production have the greatest likelihood of aiding weeds or harming nontarget species [1]. However, these outcomes seem unlikely for most currently grown transgenic crops. Many transgenic traits are likely to be innocuous from an environmental standpoint, and some could lead to more sustainable agricultural practices.

RISK ASSESSMENT

Risk is ubiquitous and unavoidable. To a great extent, therefore, our modus operandi involves assessment and management of risk. Directly observable risks are assessed and managed through heuristic processes. This direct observation may sometimes be insufficient to establish the nature and extent of risk. In such cases, we rely on other institutions, especially reputation and the rule of law [35]. Biosafety issues pertaining to the marketing of GMOs have received increasing attention by national and international agencies and regulatory bodies worldwide [2, 4, 36]. These are based on a common set of principles built on the accumulation of experience and scientific knowledge over the past decades.

Risk assessment intends to quantify risks and evaluate the probabilities of possible outcomes on the basis of scientific data. It is a fundamental part of improving quality, being the quality of products or the quality of life, and plays a central role in the innovation required to maximize benefits. A critical step in risk assessment is identification of circumstances that may give rise to an adverse effect(s) (risk identification or "what could go wrong" step) [25]. The level of risk is then estimated from both the likelihood ("how likely is it to happen" step) and severity/consequences ("would it be a problem" step) associated with the circumstances of concern. This is then followed by characterization of the risk based on evaluation of likelihood and consequences of the identified adverse effects being realized ("what is the risk" step) [11].

The international agreements such as Cartagena Protocol on Biosafety, Convention on Biological Diversity (CBD), and the International Plant Protection Convention (IPPC) address the environmental aspects of GMOs. The Article 15 of the Cartagena Protocol on Biosafety (CBD 2000) implies risk assessment to be in compliance with criteria of science and transparency using already existing and recognized techniques. The characterization process should adopt a multidisciplinary approach that

(i) analyses methodologies in statistics,

(ii) considers the individual components employed to produce the GMOs (such as characteristics of the donor organism, vector, and inserted DNA),

(iii) evaluates the final result in its totality (characteristics of the organism with new traits, information related to intended use, and characteristics of the potential receiving environment),

(iv) considers relevant information produced from both public and private research institutes and from international agencies.

The Cartagena Protocol on Biosafety in the year 2006 introduced an Annex III in the protocol of Article 15 for scientifically sound and transparent risk assessment taking into account risk assessment techniques. Such risk assessments shall be based at a minimum, on information provided in Article 8, and other available scientific evidence in order to identify and evaluate the possible adverse effects on human health and environment. The principles and methodology described in Annex III of the protocol follows the proven, well-accepted risk assessment paradigm, including identification of potential harmful characteristics of modified organisms that may have an adverse effect. Risk are then to be evaluated based on a combined analysis of the likelihood of the identified risks materializing and their consequences. The general principle of this protocol includes the following:

(i) lack of scientific knowledge or scientific consensus should not necessarily be interpreted as indicating a particular level of risk, an absence of risk, or an acceptable risk;

(ii) risks should be considered in the context of risks posed by the nonmodified recipients or parental organisms;

(iii) risks should be assessed on a case-by-base basis.

In addition, the Cartagena Protocol on Biosafety evaluated the effectiveness of the protocol (COP-MOP) for risk assessment in the Article 35 in the year of 2008 for the safe transfer, handling, and use of living modified organisms (LMO) to protect the significant loss of biological diversity.

As in the year 2010, biodiversity target is approaching to revise the current Strategic Plan of the Convention and adopt a new biosafety strategic plan (2011–2020) to implementation of more effective risk assessment strategic plans to protect the risk of GMO/LMO according to the Bureau of COP-MOP/4.

Several pieces of information would be necessary for successful risk assessment prior to release of the GMO (1) molecular characteristics of the GMO with detailed information on genetic changes in the size and sequence, (2) details of the technology used to effect the genetic changes, (3) details of the genes and their properties that have been introduced and the possible effects of any other genetic change brought about in the organism, (4) automated karyotyping and chromosomal analysis, (5) growth characteristics of the GMO in comparison with the host organism, (6) nutrient, soil, climatic, and other requirements, (7) nature of interaction with other organisms, (8) nutritional, allergenicity and toxicity studies in case of products intended for use as food and feed, (9) gene flows from the GMO under normal ecological conditions and its impact on ecology in controlled field trials, and (10) the viability of hybrids plants, its biomass productivity, and chemical composition [37].

In order to assess whether a GMO will be safe for environment, most GMOs can reproduce, multiply, and spread in the environment after they are released. The genetic modification could give GM plants, animals, or microorganisms an advantage that would allow them to increase in numbers and spread in the environment. The environmental risks from GMOs will vary, depending on the characteristics of, and the interactions among, the organism, the trait introduced through the gene, and the environment. The novelty of GMOs, the fact that like all plants they will continue to reproduce after release, the complexity of natural environments and ecosystem processes, and the unknown evolutionary fate of inserted genes, all need to be considered in predicting environmental impacts. Despite the fact that the genes being transferred occur naturally in other species, there are unknown consequences which could be the change in the organism's/plants metabolism, growth rate, and/or response to external environmental factors. These consequences influence not only the GMO itself, but also the natural environment in which that organism is allowed to proliferate.

In 1986, a publication by the Organization for Economic Cooperation and Development (OECD), called "Recombinant DNA Safety Considerations," became the first intergovernmental document to address issues surrounding the use of GMOs. This document recommended that environmental risk assessments can be performed on a case-by-case basis. Since then, the case-by-case approaches of risk assessment for GMO have been widely accepted. However, the USA. has generally taken a product-based approach of risk assessment and Europe adopted a process-based risk assessment approach for GMO [38].

Environmental Risk Assessment

Environmental risk assessment (ERA) considers the impact of introducing a GM plant into a given environment. The ERA is concerned with evaluating the potential for harm to ecosystem components given that there is exposure to the GM plant. Importantly, the focus and degree of emphasis on elements of the ERA will change during the development process for the GM plant as the scope of environmental release ranges from confined field trials of limited extent through to larger-scale trials and seed increases in more environments, and to the final unconfined commercial release.

The risk of GMO toward the environment is conducted on a case-by-case basis, is comparative, and uses lines of evidence to arrive at a holistic understanding of the nature and degree of risk posed by the particular type of environmental release being analyzed [39]. In addition, a stepwise or tiered approach of data generation and analysis is used in order that the focus be directed to consequential concerns within the universe of possibilities.

Because the universe of possible concerns relevant to ERA is very large, the process of problem formulation is especially critical in order that the risk assessment be properly framed and conducted [23, 40, 41]. The universe of concerns generally need to be addressed with a few very specific questions within context to release most of the GMOs in the environments with special references to genetically modified plants. Does the genetic modification of the plant cause it to have attributes commonly associated with weeds in managed environments? Invasiveness in natural environments? Will the transgenic element in the GM plant move into native plant populations, and so what if it does? That is, will gene flow cause a native plant to become weedy or invasive (or more so)? Or will isolated populations become extinct through hybridization with the GM plant (gene swamping)? And will the GM plant adversely impact nontarget organisms that may be of special interest because they are beneficial, endangered, threatened, or charismatic? [12]. Problem formulation is a formal process whereby the risk assessor determines relevant considerations for risk assessment from this wide host of possible concerns.

The commercial development of a GM plant proceeds in a stepwise fashion, and environmental release in the first instance is in the form of field trials that are limited in number, size, and environments in which they occur. Finally, with commercialization, the GM plant is widely deployed with little concern for its confinement. Obviously, the nature of environmental impacts that need to be addressed, and therefore the data intensity and degree of scrutiny given these impacts in the ERA, will vary with the stage of development and scale of deployment being considered [42]. In view of this, the environmental risk assessment proceeds in a tiered fashion where the problem formulation considers the specific questions to be addressed and arrives at relevant data, and data synthesis needed to undertake the appropriate ERA [43]. Therefore, the ERA is dynamic with respect to the questions addressed, the data synthesized, and the comprehensiveness of the analysis conducted.

As the environmental risk assessment iterates through tiers, conservatism in conduct and interpretation of findings is balanced against uncertainties in the state of understanding. Thus, lower-tier ERA will be highly conservative to balance uncertainty, and as higher tiers of assessment are needed, increased understanding allows for more realistic (less conservative) appraisals [44].

Risk assessment also focused on the change brought about by genetic engineering allows for detailed consideration of the potential consequences of the change relative to the way the GM plant is intended to be used and the environments in which it may be found. One of the examples of safety for food use of corn engineered to be resistant to insect pests; the risk analyst at this stage can ask how specifically the change manifested by genetic transformation/expression of an active protein conferring insect resistance to the plant and may pose risks to consumers of food derived from the GM corn. In terms of potential genetically modified food safety, key considerations are how the change may result in toxicity or allergenicity.

RISK MANAGEMENT

Once a risk is assessed, it must be managed. The management of risk is an exclusively political action, resulting in a decision regarding whether to accept or not the risk previously estimated. It can take additional aspects (e.g., socioeconomic or ethical) into consideration and concerns methods used to reduce the scientifically identified risk. Many frameworks of risk assessment methodology separate risk assessment from risk management. Some frameworks, however, consider only certain aspects of risk management (e.g., monitoring) as separate from risk assessment but other aspects of risk management (e.g., consideration of risk mitigation options) to be part of risk assessment methodology, since a final characterization of risks must take into account the effects of any mitigation options that reduce risks. The important aspect is, of course, the iterative and interlinked relationship between risk assessment and risk management [3].

Often decisions are made with incomplete information, and this leads to uncertainty. This uncertainty needs to be handled to assess the impact it might have on a decision. Biosafety regulatory frameworks should serve as mechanisms for ensuring the safe use of biotechnology products without imposing unintended constraints to technology transfer.

The Article 16 of the Cartagena Protocol of Biosafety is purely relevant to risk management of GMO. The protocol establishes and maintains appropriate mechanisms and measures strategies to regulate, manage, and control risks identified in the provision of risk assessment.

(i) The potential to harmonize national regulatory frameworks thus ensures appropriate biosafety decision making based on scientific risk assessment. If properly implemented, the protocol has the potential to encourage innovation, development, technology transfer, and capacity building in relation to biotechnology, while also achieving the goals of conservation, sustainable agriculture, and equitable sharing of the technology's benefits.

(ii) To realize its potential, however, decisions concerning protocol implementation must be carefully considered and should not place undue burdens on a technology that possesses such great potential to contribute positively to sustainable agriculture and development throughout the world.

(iii) A first-things-first approach where initial efforts focus on bringing all parties to the protocol into compliance with it as quickly as possible. Developing further requirements or fine-tuning obligations at this stage only worsens the degree of noncompliance already in existence.

(iv) Therefore, capacity building should remain the primary area of focus under the Biosafety Protocol to ensure the safe adoption of this technology. In this regard, material exists to help national governments.

(v) The users and developers of agricultural biotechnology embrace their share of the duty in the protocol implementation process and will continue to campaign for fair, science-based regulations and assist with and contribute to effective capacity building.

Risk management process also forms a second focus of the economic/political component of the GMO biosafety issue. Whereas a risk/benefit analysis concludes that risks exist with regard to a GMO introduction or other activity, but are sufficiently outweighed by the benefits of that action, it will probably still be required both practically and legally to take steps to manage the risk and to ensure that damage will be minimized [45]. Elements of currently used and proposed risk management process include a variety of different kinds of activities. To a large extent, the specific protective measures imposed on the GMO user will be determined based on scientific factors linked to specific details of the GMO and the proposed use. These issues, too, turn on the ability of the decision maker to rely on unbiased scientific experts who are able to analyze each proposal or application and determine what controls are needed, and what the best available technologies and practices are [46].

The three important components was design for risk management. These components are impact assessment, public awareness/ participation, and the design of regulatory systems. These concepts, all very important in this field, are critically important for GMO-related governance. It is not possible to overstate the importance of the public's contribution to effective decision making, as well as the importance of public awareness, within the context of government decisions on matters and activities affecting the environment [47].

Role of Impact Assessment Processes

Within the concept of risk management, the mechanism of impact assessment plays a crucial role. Although extending well beyond the scope and detail of many environmental impact assessment (EIA) procedures, the assessments mandated under national biosafety-related legislation, and especially under the Cartagena Protocol, provide a clear foundation on which at least some of a country's various decision making, permitting, labeling, and other processes relating to GMOs could be based. Unfortunately, although the need for risk assessment is undisputed, the particular parameters of that investigation are difficult to quantify in the biosafety area, given the fact that GMO introductions are a relatively new innovation. In this connection, it is important to note that the development of agreed risk management measures would provide a real benefit for both the GMOs proponents, the communities, and the ecosystems that would be most affected by the identified risks.

Public Awareness/Access to Information

Public access to information is an important cornerstone of public participation and is one of the tools that could help to realize the benefits and avoid the risks of modern biotechnology. This concept is well recognized in Principle 10 of the Rio Declaration, and in the recently adopted Åarhus Ćonvention on access to information, public participation in decision making, and access to justice in environmental matters [12].

Transparency and Capacity

Simple "transparency"and "access" to relevant documents, however, may not be sufficient in the case of biosafety issues. Arguably, the concept of access to information must include, in some way, access to the tools and expertise with which to understand that information. While merely providing access to the data will be sufficient in many developed countries that are home to highly specialized and active NGOs, even here the balance of expertise weighs heavily on the side of the GMO proponents, often the companies or institutions that developed the GMOs [48].

Labelling, Standards, and Certification

Beyond the public's access to governmental documents and processes, however, there are other mechanisms by which public awareness and access to information can be encouraged, including product labeling, food safety standards, and general consumer protection laws, all of which are designed to foster awareness and communicate public preferences to the commercial proponents of GMOs in a way that will get their attention. These mechanisms can be effective if they are accurate, specific, and clearly expressed in understandable language, unbiased, and based on full disclosure of the relevant facts by the GMO proponents. In California, a major referendum requiring disclosures of toxic and carcinogenic substances in public places and consumer goods was basically invalidated by regulations that allowed those disclosures to be made in generic terms.

Confidential Information

One of the key concerns in this regard relates to the proponent's need to maintain some information as "confidential." While the basic realities of modern business clearly underscore the need for confidentiality, it is also true that confidentiality provisions are often used as a means of avoiding disclosures. In the face of increasing recognition that activities, including especially species introduction, in one country may have serious impacts on neighbouring countries, labelling and other access to information is increasingly addressed at international and regional levels [49].

Direct Public Participation and Awareness Mechanisms

With regard to direct public participation in biosafety-related decision making, a small number of countries, including Denmark, The Netherlands, and New Zealand, are also taking a leading role in developing mechanisms for public awareness. These countries' legislative provisions require relatively broad-based stakeholder processes addressing certain aspects of modern biotechnology, including the release of GMOs. Such processes help the governments and regulatory agencies to gauge public opinion, generate dialogue, gather useful information, and develop awareness within their populations on modern biotechnology [50].

Design of Regulatory Systems for GMO Development and Use

In many different fields of endeavour, technological capacity to act has moved significantly faster than has the governmental ability to oversee and regulate it. As a consequence, many concerns relating to the risk of GMOs are directed more closely to the apparent lack of societal and governmental restraints on GMO developers and users, rather than to addressing particular scientific issues. This suggests that a third key element of the risk-management process involves a reconsideration of regulatory mechanisms and systems for governmental oversight of GMO development and use [51].

Sociocultural Impacts

It is in the area of sociocultural impacts that the controversy over GMOs and biosafety takes on its most complex aspect. On one hand, food production, food security, and livelihood improvement are all critical elements of sustainable development, to which GMOs and other products of modern biotechnology are often cited as important contributions. On the other hand, the introduction of GMOs can affect humans, (as well as animals and ecosystems), particularly at the community level, in many ways beyond direct physical sustenance, not all of which are beneficial [52]. The environmental questions surrounding biotechnology need to be addressed, yet the technology as a whole offers great promise of environmental, social, and economic benefits that should not be inhibited unnecessarily. A number of concerns should be addressed through sociocultural assessment of the impact (sociocultural risks and benefits) of GMOs. These include the biodiversity impacts of extending GMO introductions into marginal areas (which are often centres of diversity not only of wild species but also of traditional agricultural species) and into protected areas and their buffer zones [53].

WAYS TO MANAGE RISKS

Management and mitigation of risk allows feedback for validation of the initial assessment. Risks can vary depending on several factors including nature of the GMO, its intended use, and the environment receiving the GMO. Therefore, they should be assessed and managed on a case-by-case basis. The purpose of case-by-case practice is to treat every release as unique, since every GMO represents different genetic characteristics. Prior consent from the authorities is advisable in order to perform deliberate release, and field trials before the GMO may be commercialized [54, 55]. This is particularly important for genetically engineered microorganisms that have the potential to survive, persist, and spread in the environment to which they may gain access. As cited by the European food safety authority [13], the following points should be addressed when appropriate:

(i) the potential for survival and persistence in the receiving environment and any selective advantage that may be offered: in case of selective advantage, its nature should be identified along with any potential for negative effects;

(ii) the potential for gene transfer;

(iii) the potential for negative effects or consequences based on interactions with indigenous microorganisms;

(iv) possible effects on humans, animals, and plants;

(v) possible effects or (nonreversible) perturbations on biogeochemical processes.

These points may be assessed by a combination of laboratory studies, micro- and mesocosm experiments, and small-scale field releases to identify hazards and to quantify actual levels of exposure [56]. However, extrapolation of assessment from one context to another, that is, from laboratory research to small-scale field trials and finally to commercial scale is not recommended. Small-scale trials involve a lesser number of GMOs and may provide valuable information related to concerns like survival and persistence, competitive fitness, and some ecological implications of release. Commercial release, on the other hand, involves a higher number of GMOs to be released in different, complex ecosystems and needs to be carefully carried out over time and at different sites to reveal impact on relationships between species and ecosystem interactions [57].

Generally, potential risks by use of GMOs can be mitigated using risk-management strategies that may make some proposed activities acceptable. This can be achieved, for example, using confinement strategies and monitoring.

Contained Use of GMOs

The term "contained use" covers any activity involving GMOs in which measures are taken to limit contact between them and people or the environment. It relates to the actual process of genetic modification, and also to the use, storage, transport, and destruction of GMOs. Containment of GMOs can be physical or biological. Physical containment includes barriers designed to prevent organisms from escaping the laboratory and be accidentally released. This may involve the use of specially built laboratories, sterilization procedures, restriction of access, and so forth. Biological containment involves designing the organism in such a manner that they cannot grow out of the laboratory.

Monitoring the Fate of GMOs Released in the Environment

An important aspect in the process of GMO risk management is the postrelease monitoring phase initiating from the moment of environmental release. Regular inspections are necessary for effective monitoring, using a diversity of analyses over an extensive period of time. Monitoring will have two focuses: (1) possible effects of the GMO identified in the formal risk assessment procedure and (2) identification of the occurrence of adverse unforeseen effects of the GMO or its use that were not anticipated in the environmental risk assessment.

The establishment of monitoring procedures may be difficult, but such monitoring is essential not only to understand the effectiveness of the introduced organism but also to detect unexpected spread. In general, the procedures involve development or application of already existing techniques for identifying the organism in environmental samples. These procedures have been developed and are, in most cases, well accepted [58].

The scientific knowledge and experiences gained from monitoring will in turn inform the risk assessment process. Thus, the results of monitoring provide opportunities to update risk assessment continually in the light of any new knowledge.

Biosafety regulatory frameworks of GMO should serve as mechanisms for ensuring the safe use of biotechnology products without imposing unintended constraints to technology transfer. To be able to judge the sensitive balance between these aspects of GMO risk management, measuring the costs of biosafety regulation and the potential impacts on biotechnology research and development is crucial. A necessary first step to answering questions about the causes and consequences of the process of regulatory approval for new biotech crops is to understand the operation of the regulatory system and the size and structure of the costs of compliance. It seems that the compliance costs incurred by biotechnology developers are quite high, and the regulatory burden of novel biotech crops might be out of balance. Reflections on trends, challenges, and issues on risk assessment and management in a developing country context were presented. Biosafety regulatory frameworks were reviewed in relation to the development process, challenges, and trends in its formulation, especially in the context of risk assessment and management. The choice of a biosafety regime in the context of developing countries is influenced not only by the science-based approach in risk analysis but also by the social, political, and environmental governance mechanisms and experience gained in relation to practice and conventions within a particular country.

The regulatory systems designed to deal with GMO should reduce the amount of risk and create the social adaptive capacity necessary to cope with the risks associated with new technologies. There are many different ways to achieve these goals. However, the three separate methods for addressing these challenges are biosafety protocols, a moratorium, and insurance.

Currently, Austria, UK, and Germany have moratoriums, while the EU has a de facto moratorium effected from June, 1999 until 2003. Such moratoriums delay the introduction of GMO that could reduce the amount of ecological degradation produced by GM. However, moratoriums offer a number of benefits. A delay could provide the opportunity to develop institutions to effectively evaluate and monitor GMO. It would also allow science to better assess the potential indirect impacts of existing GM, such as the evolution of Bt resistance. Furthermore, a moratorium may provide the time needed to allow a richer public debate to address how to fairly balance the risks and benefits of GMO. Given the uncertainty surrounding both the likelihood and degree of potential impacts of GMO risks, it is sensible for society to purchase insurance against these risks [59]. However, due to the unknown and variable nature of risks, private insurance is virtually impossible, which forces the public to play this role. Taxes on the use of transgenic products could function as a type of social insurance, as long as such a tax was invested in ecological conservation and restoration, to mitigate against any disruption caused by GMO.

PRECAUTIONARY APPROACH

When there is a risk, two or more outcomes are possible, which one will occur is unknown but at least one of them is undesired. It is within this context that the precautionary approach from Principle 15 of the Rio Declaration [60] can be introduced.

Precautionary approach is premised on the notion of reducing, if not eliminating, risks to human health and the environment. It acknowledges the complexity and variability of the natural environment and embodies certain humility about scientific procedures and knowledge. It prioritizes the rights of those who stand to be affected by an activity, rather than those who stand to benefit from it [57]. It involves scrutiny of all available alternatives and an examination of justifications and benefits as well as risks and costs. In short, a precautionary approach involves the adoption of long-term, holistic, and inclusive perspectives in environmental protection [46, 61, 62].

Precautionary approach or precautionary principle is the borderline between science and policy or science and governance, in modern parlance. It is often divided into three components: (1) the lack of scientific certainty, (2) a risk of irreversible or serious damage, and (3) an obligation for states to take measures accordingly. The precautionary principle aims to replace uncertainty to ensure safety until other measures or solutions can be implemented. Within a policy that strives to achieve sustainable development over a long period, the precautionary principle seems to be indispensable. As the precautionary principle intends to protect the environment beyond current scientific knowledge, its implementation may not be justified or questioned on the basis of current scientific data alone.

The virtue of precautionary principle is the avoidance of risks that are impossible to assess. Its vice is that these risks, which may not even exist, can only be avoided by refusing to improve quality, being product quality or the quality of life. It has been justified by academics and pressure groups for imposing restrictions on the use of genetic modification technology claiming that the lost species and ecosystems can never be recovered for future generations. On the other hand, the development-focused environmentalists note that future generations may not come into being to appreciate these ecosystems without effective action on development imperatives.

The precautionary calculus often overlooks the fact that even when technologies introduce new risks, very often they confer net benefits—that is, their use reduces many other serious and costly hazards. For example, the use of GMOs with enhanced pest and disease resistance has reduced the use of pesticides, reducing runoff into waterways, and the exposure of workers who manufacture, transport, and apply these chemicals. It has also permitted farmers to adopt environment-friendly, no-till farming practices. Rice varieties enhanced with provitamin A and iron could drastically improve health of hundreds of millions of the malnourished in developing countries. Such tangible environmental and health benefits are usually given little or no weight in precautionary risk calculations [63].

To gain more clarity on whether the precautionary principle could be used as an effective tool to reduce the overall risks or to merely regulate risks, we can consider the experience of the European Union and the US, respectively, with the genetically modified crops [64]. The EU has ensured low rates of adoption of GM crops within Europe, but it does not follow that this has reduced overall environmental or health risks, the very reason for the precautionary approach. But the US approach, which incorporates a moderate version of the principle (requiring governmental approval prior to their commercial cultivation), has led to rapid adoption of GM crops and brought significant environmental benefits (relative to conventional crops) such as higher yields, lower pesticide usage, and increases in biodiversity [65]. Thus, while the EU's version of the precautionary principle has effectively limited GM crops by prolonging riskier practices, it has been environmentally counterproductive.

Even though several countries have adopted well-drafted environmental and biodiversity laws, reference to "precautionary principle" is missing. For instance, Malaysia's national biodiversity policy makes explicit reference to the convention on biological diversity (1992) but refrains from using the term "precautionary principle." Similarly, other countries in the region, Vietnam, Indonesia and Lao PDR, also do not directly invoke precautionary principle in their laws.

On the other hand, in several countries (e.g., India and Pakistan), the highest judicial authority has cited "precautionary principle" in its judgments. Several African countries have made explicit reference to "precautionary principle" in their laws. Examples include the 1997

Mozambique environment legislation, the 1996 general environmental law of Cameroon, and South Africa's National Environmental Management Act.

Latin American countries have also incorporated 'precaution as guiding principle' in their national environmental laws. Examples include general and biodiversity-related environmental laws in Argentina, Peru, Costa Rica, and Ecuador.

The "precautionary principle" in Australia's environmental policy is deeply rooted, as reflected in the Inter-Governmental Agreement on Environment of 1992 and the Commonwealth Environment Protection and Biodiversity Conservation Act of 1999.

This "Precautionary Principle" as mentioned above is deeply entrenched in the environmental legislations of several European countries. On the other hand, in the United States of America, precaution is rarely stated explicitly in any of its laws. However, the precautionary principles are well entrenched in several protection acts such as Endangered Species Act of 1973 and the Wild Bird Conservation Act of 1992.

A report by the European Environment Agency (2002) summarizes the European experience invoking "Precautionary Principle" for environmental management which provides useful insights on what lessons can be learned about the Precautionary Principle from several case studies. These lessons include the following: (1) respond to ignorance as well as uncertainty, (2) research and monitor for early warnings, (3) search out and address "blind spots" and gaps in scientific knowledge, (4) identify and reduce interdisciplinary obstacles to learning, (5) ensure that real-world conditions are fully accounted for and systematically scrutinize and justify the claimed "pros" and "cons", (6) evaluate alternatives, (7) promote robust diverse and adaptable solutions, and (8) use "lay" and local knowledge as well as all relevant specialist expertise.

The increasing acceptance of precautionary principles has, however, been accompanied by changes in their public profile. Many European industries decreasingly view precautionary principles as acceptable risk management approaches and increasingly view them as tools for environment and health advocates. These industries 'disenchantment is fed both by decreasing control over regulatory bodies' interpretation of "precaution" and by seeing precautionary principles endorsed by individuals with very different worldviews. These proponents advocate precaution as a response to globalization and the so-called "risk society" where risks extend over time and national boundaries, without compensating many of those affected by them [63].

Some environmental groups argue that this world requires an expanded role for NGOs, representing and interpreting precautionary principles. These groups do not dismiss science but doubt its ability to resolve issues with great uncertainties and pervasive value conflicts. However, even those who acknowledge limits to conventional science may be uncomfortable with the transfer of power where NGOs made the arbiters of ambiguity [66].

Some US agencies accepted that businesses and governments may invoke precautionary principles to protect their markets from outside competition. They point to "precautionary" bans that fall heavily on US products, such as GM corn and beef raised with growth hormones [67]. The Commission of the European Communities [68] has recognized these possibilities without offering a clear resolution.

In contrast, the above precautionary principle is anultimate key issue within the sustainable development framework. New approaches are needed to take into account the risk, uncertainty, and complexity involved. There is a need to identify normative preferences, and the stakeholders perspectives and extended peer communities are needed.

HOW TO ENSURE THE ROLE OF PRECAUTIONARY MEASURES

The precautionary approach offers the public and decision makers a forceful, common-sense approach to environmental and public health problems. In order to deliver the promise implied by its name, the principle should not increase the overall risks. To ensure that a policy is truly precautionary, one should compare the risks of adopting the policy against not adopting it. It should take care of some ethical criteria to ensure that it actually reduces overall risks when outcomes are ambiguous. These include the human mortality and morbidity threat, threats from outcomes that are irreversible or persistent, the immediacy criteria wherein immediate threats must be taken care of prior to threats that could occur later, and the uncertainty criteria where threats of harm that are more certain should take precedence over less certain harms.

As formulated at the "Wingspread Conference" [69], the principle of precautionary action has 4 parts.

(1) People have a duty to take anticipatory action to prevent harm.

(2) The burden of proof of harmlessness of a new technology, process, activity, or chemical lies with the proponents, not with the general public.

(3) Before using a new technology, process, or chemical, or starting a new activity, people have an obligation to examine "a full range of alternatives" including the alternative of doing nothing.

(4) Decisions applying the precautionary principle must be "open, informed and democratic" and "must include affected parties."

In February 2, 2000, European Commission Communication (ECC) released a note on the Precautionary Principle: the "precautionary principle" applies where scientific evidence is insufficient, inconclusive, or uncertain, and preliminary scientific evaluation indicates of there are reasonable grounds for concern that the potentially dangerous effects on the environment, human, animal, or plant health EC [2000].

It is important to emphasize that although this principle operates in the context of scientific uncertainty, it is considered by its proponents to be applicable only when, on the basis of the best scientific advice available, there is good reason to believe that harmful effects might occur [44]. The precautionary principle is most often applied in the context of the impact of human actions on the environment and human health, as both involve complex systems where the consequences of actions may be unpredictable

Principle 15 of the Rio Declaration remains by far the most widely quoted version of the precautionary principle. It states that in case of "threats of serious or irreversible damage, lack of full scientific certainty shall not be used as a reason for postponing cost-effective measures to prevent environmental degradation". It also brings in the element of proportionality by stating that measures should be applied according to the capability of the states.

The Cartagena Protocol on biosafety based on the Precautionary Principle, emerged as a result of international negotiation to reduce the transboundary movement transit, handling and use of living modified organism (LMO) that may negatively impinge on the biological diversity. Three are three core elements of the Cartagena Protocol: advance information agreement (AIA), risk assessment and the precautionary principle. The idea of the for requiring AIA of LMO is that states have a right to know what is coming into their territories, and the information should be provided in time to prepare possible harm. This procedure applies only to LMOs for introduction into environment. AIA procedures are embodied in Articles 8, 9, and 10. According to these procedures, the exporting party must communicate a written request to importer prior to transfer of LMOs intended to be introduced into environment.

The risk assessment of the Article 10 is envisioned in the protocol as guideline for parties in their decisions to import LMOs. An assessment of risk will enable them to anticipate and prevent environmental harm. Risk assessment should be performed with information available to importing state in the AIA documentation. Procedures to assess risk should be conducted in a scientifically sound manner. This article also contains explicit support for the risk management and precautionary approach.

Article 11 of the Cartagena Protocol on Biosafety requires parties to comply with information requirements set in Annex II, according to which states have provided information facility in the protocol, the biosafety clearing house, a risk assessment report according to guidelines established in AnnexIII.

Article 17 stated that the unintentional transboundary movements and emergency measures of the living modified organisms that likely have significant adverse effect on the conservation and sustainable use of biological diversity take into account risk to human health.

Article 18 of the Caratagena Protocol addressing handling transport, packaging, and identification of transboundary movements of LMOs for intentional introduction into the environment must identify the organisms as LMOs. The objective of this article is to make sure that the LMOs are handled and moved safely to avoid adverse effects on biological biodiversity and human health.

The precautionary principle of the Cartagena Protocol of Biosafety in Articles 5 and 6 mainly contained the advance information agreement and the risk assessment requirement. The inclusion of precautionary principle in the agreement to some academics contains its strongest enunciation and has been regarded "operationalization" in the body of environmental treaty. The catalyst for the application of precautionary measures in this agreement is the risk assessment. If the risk assessment shows unacceptable level of risk, then the GMO in question could be opposed to introduce in the environment.

Thus, the precautionary principle is not panacea, and it will not change world overnight, but it can make a difference in the protection of human health and environment by providing the guidance to policy makers when considering threats posed by GMOs [70, 71].

CONCLUSIONS

The use of genetically modified organisms is important in order to meet increasing demands and improve existing conditions prevalent in our environment. We are at an anxious juncture where, on one hand, we are faced with unprecedented threats to human health and environment, while on the other hand we have opportunities to change the way things are done. Regulations concerning use of GMOs need a broader basis for decision. Postrelease impacts of GMOs can follow preventive and precautionary measures based on risk assessment and management. Monitoring and detection methods are vital for risk assessment and management to control the negative environmental and health impacts. The international biosafety regulatory frameworks are sufficiently stringent in order to protect against genuine ascertainable risks, as well as the ability of decision makers to discern the appropriateness of data necessary to adequately conduct a risk assessment, which all have considerable consequences. Consideration of social, economic, and ethical issues needs to be taken care of. Application of the precautionary approach provides avenues for future development and use of genetic engineering.

FUTURE PROSPECTIVE OF GMO

Regulation of GMO deals with a transscientific problem, that is, the resolution of the problems is beyond the competence of the scientific system. Public perception and acceptance are dependent on trust and whether the products or processes benefit them as citizens and consumers. To take proper accounts of uncertainties and public concern would help to capture the benefits, minimize the risk, and provide goals for future development and use of genetic engineering. Judgment about risks should not be based on the method modification (classical or modern) but on the quality of the final product. What does the GMO contain, is it safe, and not how was the GMO made? Encouragement of new monitoring and detection methods and tools is therefore vital for assessment, control of environmental, and health impacts as well as collection of ecological knowledge of relevance to future releases.

ACKNOWLEDGMENT

Dhan Prakash, Ranjana Bhatia, and Sonika Verma contributed equally for the this review article.

REFERENCES

1. C. James, "Preview: global status of commercialized Biotech/GM crops," ISAAA Brief no. 35, ISAAA, Ithaca, NY, USA, 2006.
2. L. Alberghina, L. Frontali, and P. Sensi, Proceedings of the 6th European Congress on Biotechnology, Elsevier Science Publishers, Amsterdam, The Netherlands, 1994.
3. R. A. Hill, "Conceptualizing risk assessment methodology for genetically modified organisms,"Environmental Biosafety Research, vol. 4, no. 2, pp. 67–70, 2005.
4. H. A. Kuiper and H. V. Davies, "The safe foods risk analysis framework suitable for GMOs? A case study," Food Control, vol. 21, no. 12, pp. 1662–1676, 2010.
5. K. Birkenhead, Y. P. Wang, B. Noonan, S. S. Manian, and F. O'Gara, "Characterization of Rhizobium meliloti dct genes and relationships between utilization of dicarboxylic acid and expression of nitrogen fixation genes," in Advances in Molecular Genetics of Plant-Microbe Interactions, R. Verma and D. P. Palacios, Eds., p. 834, American Phytopathology Society, St Paul Publishers, 1990.

6. A. H. Bosworth, M. K. Williams, K. A. Albrecht et al., "Alfalfa yield response to inoculation with recombinant strains of Rhizobium meliloti with an extra copy of dctABD and/or modified nifAexpression," Applied and Environmental Microbiology, vol. 60, no. 10, pp. 3815–3832, 1994.
7. V. Corich, F. Bosco, A. Giacomini, M. Basaglia, A. Squartini, and M. P. Nuti, "Fate of genetically modified Rhizobium leguminosarum biovar viciae during long-term storage of commercial inoculants,"Journal of Applied Bacteriology, vol. 81, no. 3, pp. 319–328, 1996.
8. D. Schubert, "A different perspective on GM food," Nature Biotechnology, vol. 20, no. 10, p. 969, 2002.
9. H. Heuer and K. Smalla, "Horizontal gene transfer between bacteria," Environmental Biosafety Research, vol. 6, no. 1-2, pp. 3–13, 2007.
10. J. M. Tiedje, R. K. Colwell, and Y. L. Grossman, "The planned introduction of genetically engineered organisms: ecological considerations and recommendations," Ecology, vol. 70, no. 2, pp. 298–315, 1989.
11. G. T. Tzotzos, G. P. Head, and R. Hull, "Principles of risk assessment," Genetically Modified Plants, vol. 2009, pp. 33–63, 2010.
12. C. N. Stewart Jr., H. A. Richards, and M. D. Halfhill, "Transgenic plants and biosafety: science, misconceptions and public perceptions," BioTechniques, vol. 29, no. 4, pp. 832–843, 2000.
13. European Food Safety Authority, "Guidance document of the scientific panel on genetically modified organisms for the risk assessment of genetically modified microorganisms and their derived products intended for food and feed use," EFSA Journal, vol. 374, pp. 1–115, 2006.
14. K. L. Johnson, A. F. Raybould, M. D. Hudson, and G. M. Poppy, "How does scientific risk assessment of GM crops fit within the wider risk analysis?" Trends in Plant Science, vol. 12, no. 1, pp. 1–5, 2007.
15. N. C. Ellstrand, H. C. Prentice, and J. F. Hancock, "Gene flow and introgression from domesticated plants into their wild relatives," Annual Review of Ecology and Systematics, vol. 30, pp. 539–563, 1999.

16. C. Dennis, "The brave new world of RNA," Nature, vol. 418, no. 6894, pp. 122–124, 2002.
17. W. Martin, "Mosaic bacterial chromosomes: a challenge en route to a tree of genomes," BioEssays, vol. 21, no. 2, pp. 99–104, 1999.
18. H. Ochman, J. G. Lawrence, and E. A. Grolsman, "Lateral gene transfer and the nature of bacterial innovation," Nature, vol. 405, no. 6784, pp. 299–304, 2000.
19. P. M. Bennett, C. T. Livesey, D. Nathwani, D. S. Reeves, J. R. Saunders, and R. Wise, "An assessment of the risks associated with the use of antibiotic resistance genes in genetically modified plants: report of the Working Party of the British Society for Antimicrobial Chemotherapy," Journal of Antimicrobial Chemotherapy, vol. 53, no. 3, pp. 418–431, 2004.
20. S. Yoshida, S. Maruyama, H. Nozaki, and K. Shirasu, "Horizontal gene transfer by the parasitic plantStriga hermonthica," Science, vol. 328, no. 5982, p. 1128, 2010.
21. P. Keese, "Risks from GMOs due to horizontal gene transfer," Environmental Biosafety Research, vol. 7, no. 3, pp. 123–149, 2008.
22. V. E. Prescott, P. M. Campbell, A. Moore et al., "Transgenic expression of bean α-amylase inhibitor in peas results in altered structure and immunogenicity," Journal of Agricultural and Food Chemistry, vol. 53, no. 23, pp. 9023–9030, 2005.
23. K. M. Nielsen and J. P. Townsend, "Monitoring and modeling horizontal gene transfer," Nature Biotechnology, vol. 22, no. 9, pp. 1110–1114, 2004.
24. Gene Technology Ethics Committee, "Working paper: ethical issues arising from trans-species gene transfer," 2006, http://www.ogtr.gov.au.
25. European Food Safety Authority, "Statement of the scientific panel on GMO on the safe use of the nptIIantibiotic resistance marker gene in genetically modified plants," 2007,http://www.efsa.europa.eu/etc/medialib.
26. M. Labra, C. Savini, M. Bracale et al., "Genomic changes in transgenic rice (Oryza sativa L.) plants produced by infecting calli with Agrobacterium tumefaciens," Plant Cell Reports, vol. 20, no. 4, pp. 325–330, 2001.

27. W. Craig, M. Tepfer, G. Degrassi, and D. Ripandelli, "An overview of general features of risk assessments of genetically modified crops," International Journal of Biological Sciences, vol. 5, pp. 706–726, 2008.
28. M. J. Wilkinson and C. S. Ford, "Estimating the potential for ecological harm from gene flow to crop wild relatives," Collection of Biosafety Reviews, vol. 3, pp. 42–47, 2007.
29. K. S. Oberhauser, M. D. Prysby, H. R. Mattila et al., "Temporal and spatial overlap between monarch larvae and corn pollen," Proceedings of the National Academy of Sciences of the United States of America, vol. 98, no. 21, pp. 11913–11918, 2001.
30. J. M. Pleasants, R. L. Hellmich, G. P. Dively et al., "Corn pollen deposition on milkweeds in and near cornfields," Proceedings of the National Academy of Sciences of the United States of America, vol. 98, no. 21, pp. 11919–11924, 2001.
31. G. P. Dively, R. Rose, M. K. Sears et al., "Effects on monarch butterfly larvae (Lepidoptera: Danaidae) after continuous exposure to Cry1Ab-expressing corn during anthesis," Environmental Entomology, vol. 33, no. 4, pp. 1116–1125, 2004.
32. J. D. Pidgeon, M. J. May, J. N. Perry, and G. M. Poppy, "Mitigation of indirect environmental effects of GM crops," Proceedings of the Royal Society B: Biological Sciences, vol. 274, no. 1617, pp. 1475–1479, 2007.
33. J. E. Losey, L. S. Rayor, and M. E. Carter, "Transgenic pollen harms monarch larvae," Nature, vol. 399, no. 6733, p. 214, 1999.
34. J. D. Wolt, R. K. Peterson, P. Bystrak, and T. Meade, "A screening level approach for non-target insect risk assessment: transgenic Bt corn pollen and the monarch butterfly (Lepidoptera: Danaiidae),"Environmental Entomology, vol. 32, pp. 237–247, 2003.
35. J. Morris, "The relationship between risk analysis and the precautionary principle," Toxicology, vol. 181-182, pp. 127–130, 2002.
36. M. P. Nuti, A. Squartini, and A. Giacomini, "European community regulation for the use and release of genetically modified organisms (GMOs) in the environment," in Molecular Ecology of Rhizosphere Microorganisms, F. O›Gara, D. N. Dowling, and

B. Boesten, Eds., pp. 165–173, VCH Publishers, New York, NY, USA, 1994.

37. M. R. Smith and A. König, "Environmental risk assessment for food-related substances," Food Control, vol. 21, no. 12, pp. 1588–1600, 2010.
38. Y. Devos, P. Maeseele, D. Reheul, L. Speybroeck, and D. Waele, "Ethics in the societal debate on genetically modified organisms: a (Re)quest for sense and sensibility," Journal of Agricultural and Environmental Ethics, vol. 21, no. 1, pp. 29–61, 2008.
39. European Food Safety Authority, "Guidance document of the Scientific Panel on Genetically Modified Organisms for the risk assessment of genetically modified plants and derived food and feed," 2004,.
40. D. A. Andow and C. Zwahlen, "Assessing environmental risks of transgenic plants," Ecology Letters, vol. 9, no. 2, pp. 196–214, 2006.
41. Organization of Economic Corporation and Development, "Descriptions of selected key generic terms used in chemical hazard/risk assessment, OECD series on testing and assessment number 44, ENV/JM/ MONO(2003)15, environment directorate," in Proceedings of the Joint Meeting of the Chemicals Committee and the Working Party on Chemicals, Pesticides and Biotechnology, France, 2003.
42. G. L. Lövei and S. Arpaia, "The impact of transgenic plants on natural enemies: a critical review of laboratory studies," Entomologia Experimentalis et Applicata, vol. 114, no. 1, pp. 1–14, 2005.
43. S. Hassan, "The initiative of the IOBC/WPRS working group on pesticides and beneficial organism," inEcotoxicology: Pesticides and Beneficial Organisms, P. Haskell and P. McEwan, Eds., pp. 22–27, Kluwer Academic Publishers, Dordrecht, Germany, 1998.
44. M. Morgan and M. Henrion, Uncertainty: A Guide to Dealing with Uncertainty in Quantitative Risk and Policy Analysis, Cambridge University Press, Cambridge, UK, 1990.
45. ACRE, "Managing the footprint of agriculture: towards a comparative assessment of risks and benefits for novel agricultural systems," in Proceedings of the Advisory Committee on Releases to the Environment, London, UK, 2006.

46. G. M. Poppy, "Geneflow from GM plants—towards a more quantitative risk assessment," Trends in Biotechnology, vol. 22, no. 9, pp. 436–438, 2004.

47. United Nations Environment Programme, "Governance for sustainable and human development," UNDP policy document, 1997,.

48. K. R. Hayes, P. C. Gregg, V. V. Gupta et al., "Identifying hazards in complex ecological systems. Part 3: Hierarchical Holographic Model for herbicide tolerant oilseed rape," Environmental Biosafety Research, vol. 3, no. 2, pp. 109–128, 2004.

49. World Trade Organization, "European communities—measures affecting the approval and marketing of biotech products, panel report, WT/DS291/R, WT/DS292/R and WT/DS293/R final report," 2006,.

50. C. N. Stewart Jr., "Gene flow and the risk of transgene spread," 2008,.

51. R. I. Rose, "Tier-based testing for the effects of proteinaceous insecticidal plant-incorporated protectants on nontarget arthropods in the context of regulatory risk assessments," IOBC WPRS Bull, vol. 29, pp. 143–150, 2006.

52. L. A. Kogan, "World Trade Organization biotech decision clarifies central role of science in evaluating health and environmental Risks for regulation purposes," Global Trade and Customs Journal, vol. 2, p. 149, 2007.

53. J. D. Wolt and R. K. D. Peterson, "Agricultural biotechnology and societal decision-making: the role of risk analysis," AgBioForum, vol. 3, no. 1, pp. 291–298, 2000.

54. A. I. Myhr and T. Traavik, "The precautionary principle: scientific uncertainty and omitted research in the context of GMO use and release," Journal of Agricultural and Environmental Ethics, vol. 15, no. 1, pp. 73–86, 2002.

55. O. Renn, "Precaution and ecological risk," in Encyclopedia of Ecology, vol. 4, pp. 2909–2916, 2008.

56. M. Garcia-Alonso, E. Jacobs, A. Raybould et al., "A tiered system for assessing the risk of genetically modified plants to non-target organisms," Environmental Biosafety Research, vol. 5, no. 2, pp. 57–65, 2006.

57. S. Mayer and A. Stirling, "Finding a precautionary approach to technological developments—lessons for the evaluation of GM crops," Journal of Agricultural and Environmental Ethics, vol. 15, no. 1, pp. 57–71, 2002.
58. J. K. Jansson, "Tracking genetically engineered microorganisms in nature," Current Opinion in Biotechnology, vol. 6, no. 3, pp. 275–283, 1995.
59. R. Costanza, "Visions, values, valuation, and the need for an ecological economics," BioScience, vol. 51, no. 6, pp. 459–468, 2001.
60. United Nations Environment Programme, "Rio declaration on environment and development," inProceedings of the United Nations Conference on Environment and Development, Rio de Janeiro, Brazil, June 1992.
61. E. Fisher and R. Harding, Perspectives on the precautionary principle, Federation Press, Sydney, Australia, 1999.
62. C. Raffensberger and J. Tickner, Protecting Public Health and the Environment: Implementing the Precautionary Principle, Island Press, Washington, DC, USA, 1999.
63. G. Conko, "Safety, risk and the precautionary principle: rethinking precautionary approaches to the regulation of transgenic plants," Transgenic Research, vol. 12, no. 6, pp. 639–647, 2003.
64. I. M. Goklany, The improving state of the world: why we're living longer, healthier, more comfortable lives on a cleaner planet, Cato Institute, Washington, DC, USA, 2007.
65. M. Marvier, C. McCreedy, J. Regetz, and P. Kareiva, "A meta-analysis of effects of Bt cotton and maize on nontarget invertebrates," Science, vol. 316, no. 5830, pp. 1475–1477, 2007.
66. B. Durodie, Poisonous Dummies: European Risk Regulation after BSE, European Science and Environment Forum, Cambridge, UK, 1999.
67. J. Tait, "More faust than frankenstein," Journal of Risk Research, vol. 4, pp. 175–191, 2001.
68. Commission of the European Communities, "Communication from the commission on the precautionary principle," Tech. Rep. 2.2.2000 COM, European Comission, Brussels, Belgium, 2009.

69. "Wingspread conference on the precautionary principle," 1998,
70. J. N. Herrera, "International law and GMOs: can the precautionary principle protect the biological diversity," Boletin Maxicano de Derecheo Comparado, vol. 118, pp. 97–136, 2007.
71. T. E. Nickson, "Planning environmental risk assessment for genetically modified crops: problem formulation for stress-tolerant crops," Plant Physiology, vol. 147, no. 2, pp. 494–502, 2008.

Chapter 4

The Role of Social and Decision Sciences in Communicating Uncertain Climate Risks

Nick Pidgeon[1] and Baruch Fischhoff[2]

[1]Understanding Risk Research Group, School of Psychology, Cardiff University, Cardiff CF10 3AT, UK,

[2]Department of Engineering and Public Policy, Carnegie Mellon University, Pittsburgh, Pennsylvania 15213, USA

ABSTRACT

A major challenge facing climate scientists is explaining to non-specialists the risks and uncertainties surrounding potential changes over the coming years, decades and centuries. Although there are many guidelines for climate communication, there is little empirical evidence of their efficacy, whether for dispassionately explaining the science or for persuading people to act in more sustainable ways. Moreover, climate communication faces new challenges as assessments of climate-related changes confront uncertainty more explicitly and adopt risk-based approaches to evaluating impacts. Given its critical importance, public understanding of climate science deserves the strongest possible communications science to convey the practical implications of large, complex, uncertain physical, biological and social processes. Here, we identify the communications science that is needed to meet this challenge and the ambitious, interdisciplinary initiative that its effective application to climate science requires.

INTRODUCTION

In many nations, much of the public has long recognized the potential gravity of climate change[1, 2]. Nonetheless, few citizens or political leaders understand the underlying science well enough to evaluate climate-related proposals and controversies. As a result, it is hard for political leaders to generate and sustain broad public support for ambitious climate policies[3] or for citizens to take effective personal action[4]. Conversely, it is relatively easy for a vocal, partisan minority to sow confusion, hoping to justify delay and inaction by amplifying uncertainties[5, 6]. Without basic scientific knowledge, lay people struggle to distinguish legitimate scepticism, which all sciences welcome[7], from radical scepticism, an unwillingness to accept any evidence that might disprove the claims in question. Even those who see little merit in recent attacks on climate science, such as the 2009 controversy regarding e-mails obtained from the University of East Anglia's Climatic Research Unit and posted on the Internet[8, 9], must wonder why they went so far and what they mean for the future[10] (Fig. 1)

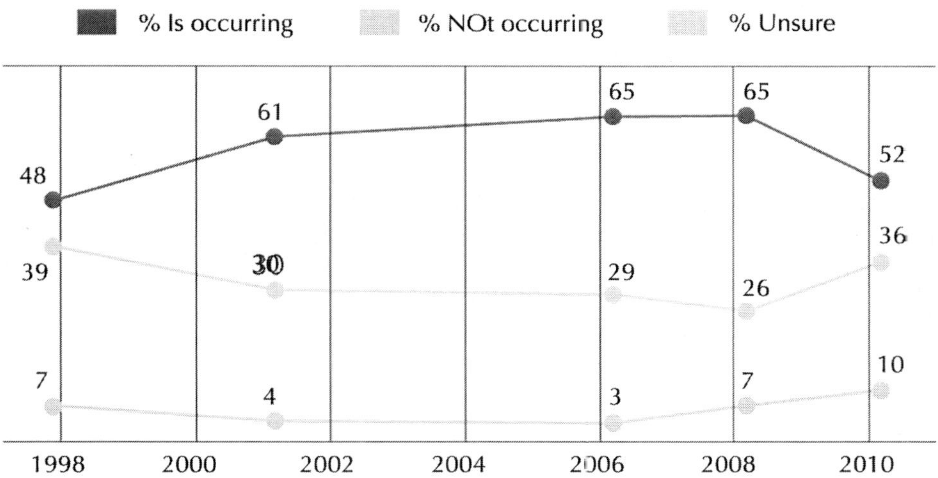

Figure 1: Shifting public opinion.

Opinion polling in America shows how people became more sure that climate scientists believed in global warming over the period 1998–2006. But more recently views are less certain, a phenomenon also seen in Britain and Europe. The reasons for this recent trend are complex and probably include a response to politicization of climate policy, as well as the impacts of the East Anglia e-mails controversy. The question asked which one of the following statements do you think is most accurate — most scientists believe that global warming is occurring, most scientists believe that global warming is not occurring, or most scientists are unsure about whether global warming is occurring or not? Data from Gallup.

These controversies may, in fact, have limited effects[11, 12]. So far, few institutions have been as trusted as science[13, 14]. The public knows scientists primarily from positive contexts, such as classrooms, news reports of breakthroughs and the many ways that science has improved their lives. Indeed, the recent stories drew such attention precisely because they depicted some scientists as violating society's expectation of independent, competent, trustworthy behaviour[15], untainted by politics. One sign of public faith in science is that even those who criticize climate science on ideological grounds use science-like language, seeking to reject the conclusions of specific scientists rather than the idea of climate science.

As painful as these attacks may be, they should not lead climate scientists to conclude that communication is hopeless. Doing so would represent what psychologists call the 'spotlight effect', exaggerating how much attention others pay to our actions. For better and worse, most people do not think about science all that much. It would be equally tragic if the attacks goaded scientists into becoming polemicists, a role that would undermine their credibility by making science seem like less trusted professions, such as the media and politics[13].

Climate scientists bear a heavy burden: potentially, the fate of the world lies partly in their hands. Fortunately, they also have some of the strongest evidence ever assembled regarding a global environmental risk. All but the most adamant critics acknowledge that global-scale climate changes have occurred before (for example, the Dansgaard–Oeschger events of the last glacial[16]) and might occur again. Where critics and mainstream scientists disagree is in their inferences about the scale and sources of the recent warming and their predictions about the impacts of continued anthropogenic climate forcing.

Realizing the practical value of climate-related research means ensuring that diverse policymakers and the public understand the risks and uncertainties relevant to the decisions that each faces. Promoting such understanding is unlikely to be a sufficient condition for individuals or societies to respond effectively to the risks posed by climate change, as a range of well-documented political[3]and psychological[4] barriers to action will also need to be addressed. However, such understanding is a necessary condition for action. In this Perspective we offer a proposal on applying the best available communications science to convey the best available climate science. The execution of this task will require sustained interdisciplinary collaboration between natural, social and decision scientists. Here we characterize the communication challenge posed by the scope, complexity and uncertainty of climate science, sketch the communications science available for this task and put forward a strategic approach to climate science communication, including the institutional support needed to sustain it.

Risk, Uncertainty and Climate Decision-Making

Uncertainty in Climate Systems

Climate scientists face many uncertainties. They progress by developing greater understanding of those uncertainties, with inquiries that sometimes reveal new sources of uncertainty[17,18]. However, their own communications about those uncertainties are very different from the ones needed by lay people and policymakers. Indeed, scientists' very familiarity with the issues can impede their ability to communicate with people outside their field, to the point that even excellent scientists can be poor communicators.

The climate system involves interactions among many moving parts, with uncertainties arising in both structuring climate models[19] and assessing their parameters and relationships[20] (Fig. 2). The systems involving climate drivers and impacts are more complicated still, and are unfamiliar to most non-scientists. Some are inherently unintuitive, such as the nonlinear relationships between emissions and atmospheric concentrations of greenhouse gases, between those concentrations and radiative forcing, and between mean global temperature changes and regional impacts. Unless scientists can convey the nature of these processes, others may over- or underestimate the definitiveness of their claims; either accepting them on faith or rejecting them outright. These uncertainties are likely to grow, as climate science extends to incorporate the socio-economic processes that drive climate changes and determine their impacts[21]. Depending on how they are communicated, these complications may further fuel fatalistic acceptance of climate-related changes (on the grounds that people and society cannot change their ways), or alternatively may highlight the fact that people, by their decisions and actions, do ultimately have the ability to avert dangerous climate change.

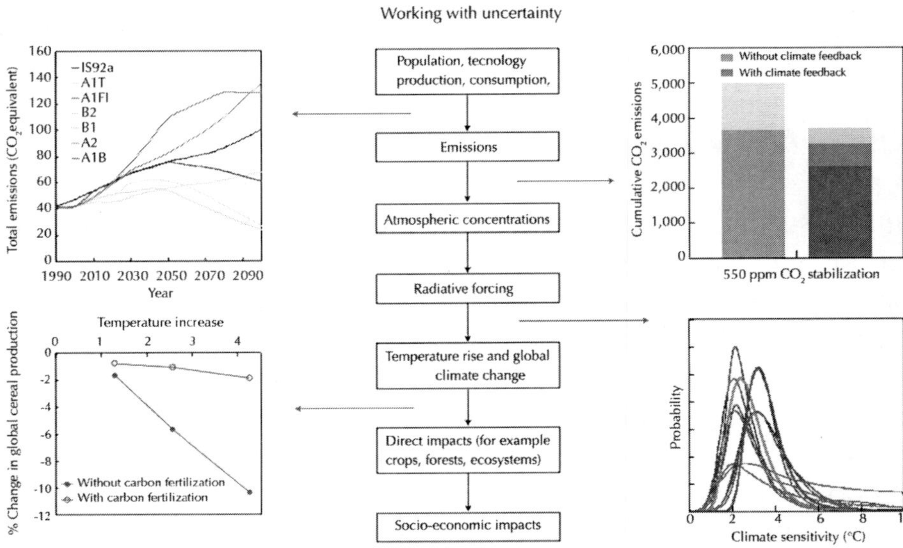

Figure 2: Sources of uncertainty in climate assessments.

The centre column shows schematically the causal chain leading from the anthropogenic sources of climate change to their impacts. The four charts show estimates for different effects in that chain, linked (by arrows) to causal factors that introduce uncertainty into these estimates. In the top-left chart, different assumptions about future human activities lead to different estimates of total emissions. The other three charts show uncertainties in cumulative emissions depending on uncertainty about climate feedback (top right), climate sensitivity depending on uncertainty about radiative forcing (bottom right), and change in global cereal production depending on uncertainty in temperature rise and carbon fertilization (bottom left). Interactions between elements in the chain also create further significant uncertainties. Reproduced with permission from ref. 27, © 2006 Cambridge Univ. Press.

In addition to its unfamiliar subject matter, much climate science relies on simulation modelling that is an unfamiliar form of inference not just for lay people but even for scientists whose disciplines use observational methods. Unless the logic of that modelling is conveyed, people may discount its conclusions. Communicating the value of climate modelling thus requires confronting such apparent contradictions as the fact that increasing a model›s complexity — by adding the behaviour of clouds, people or ecosystem feedbacks, for example — may actually increase the uncertainty in climate projections. Atmospheric scientist Kevin Trenberth of the US National Centre for Atmospheric Research in Boulder, Colorado, has explicitly warned[22] that unless such seemingly paradoxical results are communicated carefully, the more complex modelling being used in climate simulations for the upcoming fifth assessment report of the Intergovernmental Panel on Climate Change (IPCC) may confuse both the public and decision-makers, thereby reducing their willingness to act.

The discourse of scientists can be a further source of confusion. For example, scientists do not normally repeat facts that are widely accepted among them, focusing instead on the uncertainties that pose the most challenging problems. As a result, lay observers can get an exaggerated sense of scientific uncertainty and controversy, unless a special effort is made to remind them of the broad areas of scientific agreement[23]. Even that may fail unless it is made clear how 'scientific consensus' (as represented in the IPCC process for assimilating and deliberating evidence) differs from that in everyday life.

Reframing Policy through Risk and Uncertainty

The first three IPCC assessments avoided formal expressions of uncertainty, using instead scenario-based projections of possible futures. The fourth IPCC assessment[24], though, assigns explicit likelihoods (typically Bayesian assessments of probability distributions) when assessing the weight of evidence. Behavioural research has found that lay people can often extract the information that they need from clear numeric expressions of uncertainty, but that they struggle with the ambiguity of verbal quantifiers, such as 'unlikely' or 'probable'[25]. However, whether the IPCC's specific communications have succeeded is an empirical question whose answer requires rigorous research. At least one study suggests that the IPCC has not done as well as it would have wished[26].

The UK government's Stern review on the economics of climate change[27] argued for framing climate decisions in terms of the costs, risks and uncertainties of different options, as does a recent study from the US National Academies[28]. That perspective shifts the debate from whether there is anthropogenic warming to what gambles we should take with our world[29]. The former frame requires meeting an ill-defined standard of proof for demonstrating anthropogenic warming, before taking any action. The latter frame requires analysing the expected costs and benefits of different actions. The former frame leads to deliberation, the latter to decision-making.

The conclusions of these analyses ultimately depend on how decision-makers value the expected risks and benefits that are revealed[30]. For example, some critics of the Stern review argue that its discount rate was too low, relative to standard economic practice[31], thereby overweighting future outcomes. Other critics argue that its discount rate is too high, as is any non-zero discount rate when catastrophic consequences are possible[32]. Value judgements also arise in relation to many other aspects of these analyses, such as how to treat uncertain projections of the economic costs of mitigation or of the benefits of increased energy efficiency. Yet, however intense these debates, they still involve choices among options and might therefore lead to actions that avoid 'dangerous climate change', the goal enshrined in the 1992 United Nations Framework Convention on Climate Change[18, 33, 34]. In contrast, debates about evidence *per se* — such as whether there is anthropogenic climate change — can be interminable, as there will always be uncertainties to forestall decision-making.

Planning for climate adaptation necessarily requires comparison of decision options, each with its own uncertainties about the 'when, where and how' of climate impacts. For example, the UK's infrastructure planning process has 50–100-year time horizons for renewing the Thames flood-barrier scheme (Fig. 3)[35]. As a result, it must consider, among other factors, the uncertain evidence predicting that over that period the UK is likely to experience sea-level rise, more frequent summer hot spells, reduced summer rainfall in the south and east, and much wetter and stormier winter weather overall. That evidence led the UK Climate Impacts Programme scenarios for 2009 to incorporate probabilistic projections of regional impacts while acknowledging the difficulties in downscaling global projections to provide such regional forecasts (Fig. 4)[36]. In these cases, the usefulness of climate science depends on how effectively the analytical results can be communicated and how relevant they are to decision-makers.

Conceived after the 1953 flooding and associated fatalities in eastern England there are now concerns about its ability to protect London from any increased risks associated with climate change, and in particular rising sea levels and future storm surges. This has led to an extensive risk assessment and appraisal of the options for its renewal or replacement, for the period up to 2070 and beyond.

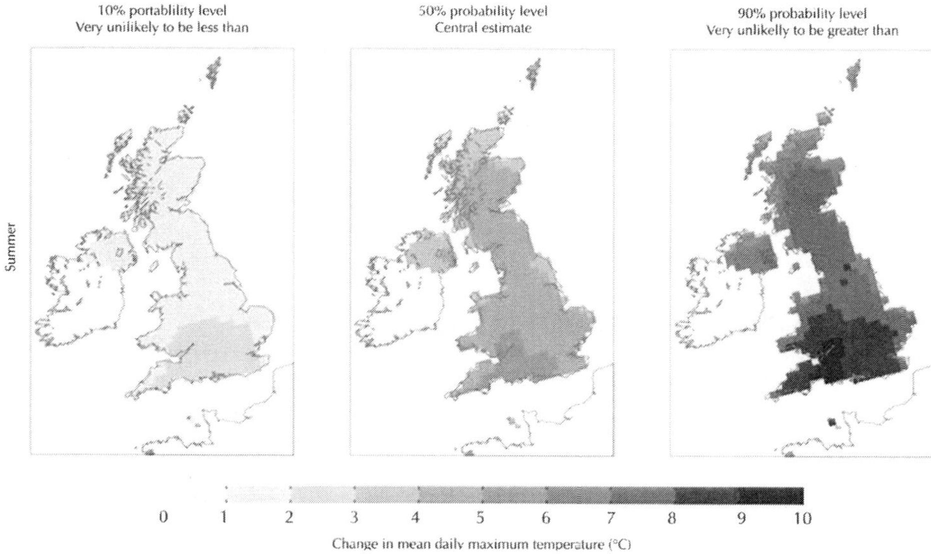

Figure 3: The existing Thames flood-protection barrier at Greenwich, East London, UK.

Figure 4: The recent UK climate projections in 2009 use probability assessments to represent the modelling uncertainty for a range of variables and with a grid resolution of 25 km^2.

The figures show 10, 50 and 90% probability levels respectively for changes to the mean daily maximum temperature in summer across the United Kingdom by 2080 and under a medium emissions scenario. Hence, within the constraints of the modelling, at the 10% level it is very unlikely (equivalent to a 10% probability) that the temperature will be less than that shown at any particular location. Reproduced with permission from ref. 70, © UK 2009 Climate Projections.

The Science of Communication

Climate scientists' struggles in communicating with the non-specialist audience are typical of the challenges that arise with any policy question having significant technical content, whether it be nuclear power[37],[38], genetically modified crops[39] or nanotechnology[40]. Over the past 40 years, social and decision science research has addressed these challenges across a broad front. This body of research has identified basic processes guiding risk perceptions and has applied that knowledge to facilitating informed choices involving complex, uncertain and contested science[41,42]. Studies focused on lay perceptions of climate change echo these general patterns with some unique specifics. For example, in western countries most people are concerned about the problem, although few make it a top priority[2, 43]. Most have fragmentary mental models that are more complete for the consequences of climate change than for its causes[44, 45]. Whatever their beliefs, most people find climate change psychologically distant, as something that will affect other people in other places and times[1, 46]. Although they generally understand the broad outlines of climate change, people can be confused by the uncertain signals sent by extreme weather events. People with different cultural values may have different beliefs[47],[48], yet still act similarly (for example, both concerned individuals and critics are more likely than disinterested people to adopt energy-saving practices[49]). In addition, many lay people ascribe primary responsibility for dealing with climate change to national governments and other powerful organizations and agents[11, 50], concluding, quite sensibly, that climate change is too big a problem for them to tackle as individuals alone. In turn, many governments fear punishment by citizens at the ballot box if they propose overly radical actions[3], leading to a collective failure to act on the part of both governments and individuals.

Research in social and decision science has identified several key lessons that are especially relevant to communicating climate science. The first is that 'risk' can be defined in different ways, by both analysts and lay people[51], depending on how they value the outcomes at stake[52]. For example, some people care primarily about threats to human life; others care about the economy or the environment as well, and need different risk estimates. When concern focuses on human life, some people care about the age of those at peril; others do not. Depending on which of these two positions decision-makers take, they will need different estimates: expected chance of premature death for those in the first group; expected life years lost for those in the second. Proponents of environmental justice need to know which groups bear the risks and which get the benefits of proposed policies. Some people need to know the extent to which risks are voluntary, controllable, uncertain, irreversible and catastrophic[53]. Unless they receive the information that they need, people must guess at it, reading between the lines of scientists' statements and controversies.

But understanding risk requires more than just knowing risk estimates. People also need cognitive representations[41] (or 'mental models') of the processes creating and controlling the risks, and thus causing uncertainty about them. For example, they may need to know how warmer oceans affect tropical storms, marine phytoplankton and winter precipitation, or how rising atmospheric carbon dioxide levels lead to increased ocean acidification. Knowledge of such processes allows people to follow public debates and grasp the rationale for alternative policies. It protects (or 'inoculates') them from being 'blind-sided' by unfamiliar facts or perspectives. It affords them the warranted feelings of self-efficacy needed before acting.

Recent advances in behavioural and decision science also tell us that emotion is an integral part of our thinking, perceptions and behaviour, and can be essential for making well-judged decisions[54,55]. Although it can cloud judgment[48, 56], emotion can provide cues valuable to evaluating evidence and the people who provide it. Emotion creates the abiding commitments needed to sustain action on difficult problems, such as climate change. It motivates climate scientists, as well as their audiences and critics[57]. Clear, respectful messages can reduce the destructive emotions of impatience, frustration and anger, and appropriately framed emotional appeals can motivate action, given the right supporting conditions[58] (in particular a sense of personal vulnerability, viable ways to act, feelings of personal control and the support of others).

Finally, social processes can amplify or attenuate perceptions of risk[59]. Sometimes those processes undermine lay understanding, as with interest groups' disinformation campaigns. Sometimes they create trust between lay and expert communities, as with well-designed public consultations. At their very best, they create effective social oversight and governance structures[60]. At their worst, they subordinate science to ideology[5, 48].

Of course, information alone will not ensure wise climate-related decisions. Social scientists have documented a range of structural, political and economic barriers to strong action affecting both institutions[3] and individuals[4]. Moreover, as noted above, for some partisans science-like arguments are just a means to political or economic ends. Well-informed individuals can rationally do nothing if they see no viable actions. Well-informed collectives can be paralysed when they realize that their members have conflicting goals. Resolving such impasses requires applying climate science to the creation of better options. Whatever the barriers might be, scientists' obligation is to provide the information that is necessary, if not sufficient, for informed choices. Having fulfilled this minimum condition, they are free to weigh in as ordinary citizens, with their own personal votes on how to make the difficult choices that climate science has identified. They might find it easier to make their case if good science communication has expanded the range of reasoned discourse.

These research results, and others like them, belie the simple behavioural theory underlying the 'deficit model' of the public understanding of science, which assumes that simply teaching more science will bring lay behaviour into line with scientists' expectations. Although the limits to this model are well documented[61, 62], it has such strong intuitive appeal that communication must explicitly adopt an alternative strategy if it is to respect audiences' values, feelings and need for dialogue and engagement[42, 63]. Here we offer such a strategy.

A Strategic Approach to Climate Communication

Strategic Listening

Climate science has always taken a long-term integrated approach. The communication of that science must be just as strategic in its analysis, design, implementation and evaluation. Like climate systems, human systems involve the complex, uncertain interaction of many processes. As a result, even the best-designed communications, based on the strongest social and decision science, require rigorous implementation and empirical evaluation to determine how effective they are. Yet, despite the critical importance of climate-change communication, such evaluations are remarkably rare. Instead, most communications rely on intuitive notions of what to say and how to say it.

Unfortunately, people often have flawed intuitions regarding how well they communicate, typically exaggerating their success. In ordinary conversation, people receive feedback, allowing them to refine imperfect communications. Scientists, though, often have little direct contact with the public. As a result, they cannot tell how well they are doing or how to do better. Without evidence to moor them, science communication can lurch from one well-intended initiative to the next. A scientific approach to communicating science requires the systematic feedback provided by empirical evaluation.

In this strategy, social and decision science research provides connections that scientists normally lack. Its methods (for example surveys, interviews and moderated deliberations) can be seen as surrogate conversations, conducted by researchers on behalf of climate scientists, to establish which outcomes matter to different people, which options they have for attempting to realize their goals and what additional knowledge they need[63, 64]. The decision science approach takes an 'inside view', letting decision-makers' needs determine the content of communications, rather than just relaying the messages that scientists think are important. With some decisions, research might find that people already know enough about the relevant climate science, but need to know more about risks to prized ecosystems or about the viability of cap-and-trade schemes. Research might find that people know enough about the facts of a decision, but are uncertain how to make the hard trade-offs that it requires (for example between risks to the present generation and to future ones). If so, then people do not need more facts but more perspectives, helping them to think about what the decision means to them and the outcomes that they value. There is no way to know what information people need without doing research that begins by listening to them.

Such listening protects against simplistic solutions to complex problems. For example, if scientists conclude that uncertainty means paralysis, they may avoid discussing it. However, if lay people know that climate risks involve complex, uncertain processes, they may resent being patronized with overly certain accounts. British readers may remember the promise that British beef was completely safe to eat at a time when the evidence regarding the transfer of BSE (bovine spongiform encephalopathy) to humans was not yet settled. US readers may remember their government's assurances that flying was completely safe when it lifted the ban imposed after the September 11 attacks in 2001, even though such terrorist threats were only just beginning to be understood. With contested science, uncertainties eventually emerge. Scientists can protect themselves by being the first to acknowledge the uncertainties in their work. Those admissions need not encourage paralysis if the results are framed as illuminating the relative risks of action and inaction rather than as inviting delays until they are completely resolved.

Strategic listening also protects scientists from confusing ignorance with incompetence. When lay people misunderstand some aspect of science, it may be that they are incapable of grasping it. However, the science may also have been poorly explained or seemed too irrelevant to master. Not knowing facts that are second nature to scientists does not, by itself, provide grounds for writing off the lay public. Doing so runs the risk of souring scientists' relations with non-scientists, who will sense the lack of respect. By dismissing non-scientists, experts also forgo the chance to fill the gaps in their fragmentary mental models. For example, lay people often confuse healthy lawns with healthy ecosystems, think that turning thermostats past the set point will make homes heat up faster and believe that the ozone hole is a cause of global warming[44]. The flaws in these beliefs are not hard to explain, once research has identified them. Indeed, such research has helped to reduce the prevalence of the last of these misconceptions[2].

Strategic listening can also assess the cumulative impact of the diverse messages on a topic. For example, it could determine when a flood of uncoordinated climate-related recommendations produces inaction ("I'll wait until the experts sort things out"), a symbolic act or two ("because you can't do everything") or sweeping lifestyle changes ("I hadn't realized the scope of the problem"). It could determine when messages about adaptation discourage mitigation by making the future seem tolerable, or encourage it by making the risks more real. Or it could focus messages on the few actions most worth doing, hoping to encourage a cascade of effective actions—sometimes called 'behavioural spillover'[65].

A strategic communication perspective will find that there is no single simple recipe for climate communication[66]. At the individual level, even the strongest social and decision science produces only better best guesses about how to formulate messages. As a result, empirical testing is always needed. At the society level, people with different goals and knowledge will have different needs to hear and be heard. For example, energy conservation programmes that motivate some people fail with others who hold similar political values, while motivating people with different values[49]. Changes in the science, the economy and politics may bring different issues to the fore, requiring communications to evolve. As a result, continued listening is needed.

Strategic Organization

Communications worthy of climate change will require sustained contributions from cross-disciplinary teams, working within an institutional framework that provides support for their efforts. Such teams would include, at minimum, climate and other experts, decision scientists, social and communications specialists, and programme designers (Box 1). Once assembled, these teams must be coordinated so that experts stay focused on their aspect of the communication process. For example, subject-matter experts should edit for fact, not style; they should also check that social scientists have not garbled the facts when trying to make them clearer. That coordination must maintain a rhetorical stance of non-persuasive communication[67], trusting the evidence to speak for itself, without spin or colouring. Although there is an important place for persuasive communication, encouraging individual behaviours and public policies, it must be distinct, lest scientists come to be seen as inept politicians. If climate scientists passionately offer dispassionate accounts of the evidence, it will preserve their uniquely trusted social position and avoid the advocacy that most are ill-suited to pursue by disposition, experience and training.

Box 1: Expertise Needed For Effective Communication of Climate Science to Aid Climate-Related Decision-Making

Subject-matter experts who can represent the latest science on topics that matter to their audiences. Those scientists migh include climatologists, ecologists, economists, engineers, physicists or others, as the decisions require. Decision scientists who can identify the most relevant aspects of that science and summarize it concisely, including its uncertainties and controversies.
Social and communications scientists who can assess the public's beliefs and values, propose evidence-based designs for communication content and processes, and evaluate their performance.
Programme designers who can orchestrate the process, so that mutually respectful consultations occur, messages are properly delivered and policymakers hear their various publics.

Creating such teams will require an organization with unique capabilities, as yet absent in the climate change arena. It must nurture the sustained interactions needed to build trust among disciplines unaccustomed to collaboration. It must maintain public trust by faithfully representing the science and its attendant uncertainties and controversies. Such a commitment to candour is routinely found in academic institutions, but often without a core commitment to practical and collaborative research. Transcending academic norms requires 'boundary organizations', chartered both to conduct basic science and to translate its results into decision-relevant terms[68, 69]. It requires the resources to support research, design and evaluation services for all scientists hoping to communicate their work. It requires the stature needed for those scientists to value its services. And it requires the staying power needed to develop new research methods and career paths, allowing its collaborators to pursue the basic research topics that it identifies. Having such capabilities should reduce the risk of unwarranted controversies and provide evidence-based responses to them.

Models for such interdisciplinary 'big science' might include the RAND Corporation (in the US) and the International Institute for Applied Systems Analysis (in Austria). In the UK, the Tyndall Centre for Climate Change Research has fostered a ten-year collaboration between climate and social scientists, although without a major focus on communication and decision-making research. In the US, the National Center for Atmospheric Research has made similar efforts. Each of these institutions performs publication-quality research on applied problems, using analytical methods to integrate diverse data, while contributing critical expertise to address policy problems. Each provides an environment that attracts scientists for individual projects, short stays and careers. Each has leadership willing and able to redeploy resources from over-represented disciplines to under-represented ones. Although smaller, more distributed models might be envisaged[28], the science of communicating science has become so important that it requires equivalent institutions, properly funded and staffed.

CONCLUSIONS

Many climate scientists are understandably frustrated by the limited response to what they see as the greatest threat facing our planet. One impulsive response to a seemingly recalcitrant public is a big advertising campaign. However, unless founded on sound social and decision science principles and accompanied by rigorous empirical evaluation, such efforts have little chance of sustained success. Moreover, each communication failure makes future success less likely, by eroding both the public's trust in the experts, who seem not to know their needs, and the experts' trust in the public, which seems unable to understand the issues. Given the gravity and the complexity of climate-related decisions, we need a new model of science communication, with new collaborations among the sciences at both the national and the international level.

AUTHOR CONTRIBUTIONS

N.P. and B.F. contributed equally to the formulation and writing of the manuscript.

ACKNOWLEDGEMENTS

Work for this paper was supported by a Personal Fellowship to N.P. from the UK Economic and Social Research Council (RES-066-27-0013), with additional support from the Leverhulme Trust (F/00 407/AG) and the US National Science Foundation (SES-0949710). We wish to thank I. Hall and P. Stern for comments on earlier versions, together with attendees at a workshop held at Cumberland Lodge, Windsor, UK in the summer of 2010.

REFERENCES

1. Lorenzoni, I. & Pidgeon, N. F. Public views on climate change: European and USA perspectives. *Clim. Change* 77, 73–95 (2006).
2. Brechin, S. R. in *Routledge Handbook of Climate Change and Society* (ed. Lever Tracey, C.) 179–209 (Routledge, 2010). The most complete current analysis of public beliefs and attitudes regarding climate change, with good international coverage including the developing world.
3. Compston, H. & Bailey, I. (eds) *Turning Down the Heat: The Politics of Climate Policy in Affluent Democracies* (Palgrave Macmillan, 2008).
4. Lorenzoni, I., Nicolson-Cole, S. & Whitmarsh, L. Barriers perceived to engaging with climate change among the UK public and their policy implications. *Glob. Environ. Change* 17,445–459 (2007).
5. Dunlap, R. & McCright, A. in *Routledge Handbook of Climate Change and Society* (ed. Lever Tracey, C.) 240–259 (Routledge, 2010).

6. Oreskes, N. & Conway, E. M. *Merchants of Doubt* (Bloomsbury, 2010). A provocative historical and political analysis of attempts to create uncertainty and interpret it as reason for inaction.
7. Corner, A. Do climate change skeptics give skepticism a bad name? *Guardian Environment Blog Online* (22 February 2010); available via http://go.nature.com/MoX91n.
8. House of Commons Science and Technology Committee *The Disclosure of Climate Data from the Climatic Research Unit at the University of East Anglia*. 8th Report of Session 2009–2010, HC-387-1 (Stationary Office, 2010).
9. Nehrlich, B. 'Climategate': paradoxical metaphors and political paralysis. *Environ. Values* 19,419–442 (2010).
10. Climate of fear. *Nature* 46, 141 (2010).
11. Spence, A., Venables, D., Pidgeon, N. F., Poortinga, W. & Demski, C. *Public Perceptions of Climate Change and Energy Futures in Britain: Summary Findings of a Survey Conducted in January-March 2010*. Understanding Risk Working Paper 10–01 (Cardiff University, 2010).
12. Krosnick, J. The climate majority. *The New York Times* (10 June 2010); available viahttp://go.nature.com/uJJGSt.
13. Market & Opinion Research International *Science in Society: Findings from Qualitative and Quantitative Research* (Office of Science and Technology, 2005).
14. National Science Board *Science and Engineering Indicators* Ch. 7, 7.1–7.46 (National Science Foundation, 2006).
15. Poortinga, W. & Pidgeon, N. F. Exploring the dimensionality of trust in risk regulation. *Risk Anal.* 23, 961–972 (2003).
16. Kunzig, R. & Broecker, W. *Fixing Climate* (Profile Books, 2009).
17. Oppenheimer, M. Defining dangerous anthropogenic interference: the role of science, the limits to science. *Risk Anal.* 25, 1399–1407 (2005).
18. Schellnhuber, H. J. (ed.) *Dangerous Climate Change* (Cambridge Univ. Press, 2006).
19. National Research Council *Advancing the Science of Climate Change* (National Academies Press, 2010). An important recent statement of the US National Academies, emphasizing the integration of natural, social and decision sciences in advancing climate science and policy.

20. Bony, S. *et al*. How well do we understand and evaluate climate change feedback processes? *J. Clim*. 19, 3445–3482 (2006).
21. Pidgeon, N. F. & Butler, C. Risk analysis and climate change. *Environ. Polit*. 18, 670–688(2009).
22. Trenberth, K. More knowledge, less certainty. *Nature Rep. Clim. Change* 4, 20–21 (2010).
23. Somerville, R. C. J. How much should the public know about climate science? *Clim. Change*104, 509–514 (2011).
24. IPCC *Climate Change 2007: Fourth Assessment Report* (Cambridge Univ. Press, 2007).
25. Budescu, D. V & Wallsten, T. S. in *The Psychology of Learning and Motivation* (eds Busemeyer, J. R., Hastie, R. & Medin, D.) 275–316 (Academic, 1995).
26. Budescu, D.V., Broomell, S. B. & Por, H. Improving communication of uncertainty in the reports of the IPCC. *Psychol. Sci*. 20, 299–308 (2009).
27. Stern, N. *The Economics of Climate Change: The Stern Review* (Cambridge Univ. Press,2006).
28. Corell, R. W. & Lee, K. N. (eds) *Informing Decisions in a Changing Climate* (National Academies Press, 2009).
29. Dietz, S., Hope, C. & Patmore, N. Some economics of 'dangerous' climate change: reflections on the Stern review. *Glob. Environ. Change* 17, 311–325 (2007).
30. Fischhoff, B., Lichtenstein, S., Slovic, P., Derby, S. L. & Keeney, R. L. *Acceptable Risk*(Cambridge Univ. Press, 1981). A critical analysis of fundamental approaches to making decisions about risk, treating risks as acceptable only in the context of compensating benefits.
31. Nordhaus, W. D. A review of the Stern review of the economics of climate change. *J. Econ. Lit*. 45, 686–702 (2007).
32. Weitzmann, M. A review of the Stern review of the economics of climate change. *J. Econ. Lit*. 45, 703–724 (2007).
33. Dessai, S. *et al*. Defining and experiencing dangerous climate change. *Clim. Change* 64,11–25 (2004).
34. Lorenzoni, I., Pidgeon, N. F. & O'Connor, R. Dangerous climate change: the role for risk research. *Risk Anal*. 25, 1387–1398 (2005).

35. Estuary, T. *2100 Plan: Managing Flood Risk Through London and the Thames Estuary*(Environment Agency, 2009).
36. Department for Environment, Food and Rural Affairs *Adapting to Climate Change: UK Climate Projections* (DEFRA, 2009).
37. Pidgeon, N. F., Lorenzoni, I. & Poortinga, W. Climate change or nuclear power — no thanks! A quantitative study of public perceptions and risk framing in Britain. *Glob. Environ. Change*18, 69–85 (2008).
38. Rosa, E. A. & Clarke, D. L. in *Research in Social Problems and Public Policy* Vol. 7 (eds Youn, T. I. K. & Freudenburg, W. R.) 21–57 (JAI, 1999).
39. Shanahan, J., Brossard, D. & Nesbitt, C. (eds) *The Public, the Media and Agricultural Biotechnology* (CABI, 2007).
40. Pidgeon, N. F., Harthorn, B., Bryant, K. & Rogers-Hayden, T. Deliberating the risks of nanotechnology for energy and health applications in the US and UK. *Nature Nanotech.* 4,95–98 (2009).
41. Morgan, M. G., Fischhoff, B., Bostrom, A. & Atman, C. J. *Risk Communication: A Mental Models Approach* (Cambridge Univ. Press, 2002). This book develops a methodological approach to integrating risk analysis and risk communication, focusing analysis on decision-makers' needs and communication on critical gaps.
42. Stern, P. C. & Fineberg, H. V. (eds) *Understanding Risk: Informing Decisions in a Democratic Society* (National Academy Press, 1996). An influential report identifying the essential role of values in setting the terms of risk analysis and advocating an analytic-deliberative approach to resolving them.
43. Nisbet, M. C & Myers, T. Twenty years of public opinion about global warming. *Public Opin. Q.* 71, 1–27 (2007).
44. Bostrom, A., Morgan, M. G., Fischhoff, B. & Read, D. What do people know about global climate change? 1. Mental models. *Risk Anal.* 14, 959–970 (1994).
45. Kempton, W. Public understanding of global warming. *Soc. Nat. Resour.* 4, 331–345 (1991).
46. Bord, R. J., Fisher, A. & O'Connor, R. E. Public perceptions of global warming: United States and international perspectives. *Clim. Res.* 11, 75–84 (1998).

47. Douglas, M. & Wildavsky, A. *Risk and Culture: An Essay in the Selection of Technological Dangers* (Univ. California Press, 1982).
48. Hulme, M. *Why We Disagree About Climate Change* (Cambridge Univ. Press, 2009).
49. Leiserowitz, A., Maibach, E. & Roser-Renouf, C. *Yale Project on Climate Change: Global Warming's Six Americas, January 2010* (Yale Univ. & George Mason Univ., 2010); available at http://environment.yale.edu/climate/files/SixAmericasJan2010.pdf.
50. Bickerstaff, K., Simmons, P. & Pidgeon, N. F. Constructing responsibility for risk(s): negotiating citizen-state relationships. *Environ. Plan. A* 40, 1312–1330 (2008).
51. The Royal Society *Risk: Analysis, Perception and Management* Ch. 5, 89–134 (The Royal Society, 1992).
52. Fischhoff, B. Risk perception and communication unplugged: 20 years of process. *Risk Anal.*15, 137–145 (1995).
53. Slovic, P. *The Perception of Risk* (Earthscan, 2000). A collection of core readings in risk perception and communication.
54. Marx, S. B. *et al.* Communication and mental processes: experiential and analytic processing of uncertain climate information. *Glob. Environ. Change* 17, 47–58 (2007).
55. Slovic, P. *The Feeling of Risk* (Earthscan, 2010).
56. Janis, I. & Mann, L. *Decision Making: A Psychological Analysis of Conflict, Choice and Commitment* (Free Press, 1976).
57. Risbey, J. S. The new climate discourse: alarmist or alarming? *Glob. Environ.* 18, 26–37(2008).
58. Spence, A. & Pidgeon, N. F. Framing and communicating climate change: the effects of distance and outcome frame manipulations. *Glob. Environ. Change* 20, 656–667 (2010).
59. Pidgeon, N. F., Kasperson, R. K. & Slovic, P. (eds.). *The Social Amplification of Risk*(Cambridge Univ. Press, 2003). A collection of research into the social dimensions of risk communication.
60. Renn, O. *Risk Governance: Coping with Uncertainty in a Complex World* (Earthscan, 2008).
61. Wynne, B. & Irwin, A. *Misunderstanding Science* (Cambridge Univ. Press, 1996).

62. Sturgis, P. & Allum, and N. Science in society: re-evaluating the deficit model of public attitudes.*Public Understand. Sci.* 13, 55–74 (2004).
63. Arvai, J., Gregory, R. & McDaniels, T. Testing a structured decision-aiding approach: value-focused thinking for deliberative risk communication. *Risk Anal.* 21, 1065–1076 (2001).
64. Clemen, R. *Making Hard Decisions: An Introduction to Decision Analysis* (Duxbury, 1996).
65. Thorgerson, J. & Ölander, F. Spillover of environmentally friendly consumer behavior. *J. Environ. Psychol.* 23, 225–236 (2003).
66. Moser, S. Communicating climate change: history, challenges, process and future directions.*WIREs Clim. Change* 1, 1–27 (2010). An informative review of basic research findings on risk communication as applied to climate change.
67. Fischhoff, B. Non-persuasive communication about matters of the greatest urgency: climate change. *Environ. Sci. Technol.* 41, 7204–7208 (2007).
68. Guston, D. H. Boundary organizations in environmental policy and science: an introduction.*Sci. Technol. Hum. Values* 26, 399–408 (2001).
69. Lorenzoni, I., Jones, M. & Turnpenny, J. Climate change, human genetics, and post-normality in the UK. *Futures* 39, 65–82 (2007).
70. http://www.ukclimateprojections.defra.gov.uk.

Chapter 5

Technological Risks, Transgenic Agriculture and Alternatives

Pablo Rubén Mariconda

Department of Philosophy, Faculty of Philosophy, Letters, and Human Sciences, Institute for Advanced Studies, University of São Paulo, Brazil.ariconda@usp.br

ABSTRACT

After discussing the transformation of age-old agricultural practices that has been occurring since the mid nineteenth century, and its impact on the natural environment, I identify four features of technology that point to the ambiguity of the idea of "technological progress". These are linked to the intrinsic unpredictability (and uncertainty) of technological applications and have implications for evaluating technological risks. I then show that large scale technological applications and innovations - such as expanding the practice of smallpox inoculation in the second half of the eighteenth century - occur in states of 'technological exception'. In them, values are suspended and rules, limits and security norms are absent; and, in the face of risks, losses and harm to health and environment that are occasioned by technological applications introduced in the name of economical progress (development), normal social inhibitions become suspended in the name of furthering the values of capital and market. In addition, I show that maintaining technological exception has favored, on the one hand, the development by large corporations of antiscientific practices that threaten the moral integrity (ethos) of science by impeding the proper use of the scientific method for evaluating the consequences of using technologies that are protected by patents; and, on the other hand, under the cover of ignorance, continued deterioration of the environment that threatens the very survival of the human species.

"Indeed the risk you run in purchasing knowledge is much greater than that in buying provisions." Plato,*Protagoras,* 814 A

In this article I adopt the perspective of the model of the interaction between scientific activities and values (cf. Lacey & Mariconda, in press). I consider particularly the application of scientific knowledge, so that my argument moves between the frontiers of science and technology. I discuss the complex relations and tensions between science and technology, attending particularly to what is called post-modern technoscience (cf. Mariconda, 2012), the predominance of practical and utilitarian interests in the politics of science, technology and innovation, and the hegemony of the values of control and (economic) progress or development (cf. Mariconda, 2006; Lacey, 2010).

My argument is not complicit in what Putnam (cf. 2004, chapter 8) denounced as the "philosophers of science's evasion of values" throughout the twentieth century. Moreover, it rejects the view that values are expressions of personal and subjective preferences that are completely distinct from the objective facts (cf. Mariconda, 2006), and affirms with Dewey that "value is something that has to do with all of experience" (Dewey *apud* Putnam, 2004, p. 135). Taking values into account changes the axis of the relation between ethics and epistemology. It enables treating together the ethical problem of "the relation between personal freedom and the stable order" (Dewey apud Putnam, 2004, p. 178, note 2) and the general epistemological problem of the relations of the objective and the subjective in experience, with the consequence that it rejects confining ethics to the restricted sphere of subjectivity and so is able to recognize the proper role of ethics in scientific and technological activities. The evaluative perspective adopted is thus in line with the pragmatist outlook expressed in the formula: "the knowledge of facts presupposes the knowledge of values" (Putnam, 2004, p. 145), that abandons the empiricist dogma: "facts are objective and values are subjective and 'never the twain shall meet'" and accepts the challenge of considering scientific and technological practices from both epistemological and ethical perspectives.

In the light of this stronger connection between epistemology and ethics, I will address, on the one hand, epistemological questions about the unpredictability and uncertainty of (mainly big scale) technological applications, focusing on a major expression of commercially oriented technoscience, transgenic agriculture and its risks and impacts on health and environment (cf. Lacey, 2006, 2008) and, on the other hand, ethical questions related to the antiscientific behavior of large corporations. I will concentrate on the environmental question because the destiny of humanity is at stake in how it is dealt with (cf. Dupas, 2006, 2008).

THE TRANSFORMATIONS OF THE NATURAL ENVIRONMENT IN AGRICULTURE

The impacts of extensive mechanization and growing artificiality of the natural environment produced by technology can be seen in a striking way in the age-old practices of agriculture, which has been more and more transformed into a "technical environment", one increasingly mediated by technical objects and actions. To Friedmann, this transformation from a "natural environment" to a "technical environment" is generated by what he calls the "industrialization of agriculture", particularly with the large scale introduction of agricultural machines into the fields (furthered by the electrification and motorization), that facilitated the widespread growing of monocultures (cf. Friedmann, 1968, p. 9).

This process of mechanization corresponds to the first stage of industrialization of the agricultural field and to the profound technical transformation of the agrarian natural environment, which furthers the tendency for technoscientific societies "to transform agriculture into an industry in accordance with Ford's definition of agriculture as 'an industry producing foods' and the modern mechanical developments as 'factories in the fields'" (1968, p. 118). Friedmann considered mainly the accompanying transformation of fields into cities that is an integral part of the all encompassing industrialization process that began in the middle of nineteenth century and continued throughout the twentieth century, and that was associated with a process of conurbation,[1] provoked by accelerated urbanization of peripheral zones of cities, where traditionally— the plantations that supplied the cities were located.

Footnote

[1]The term "conurbation" is attributed by Lewis Mumford (cf. 2008 [1961], p. 559, 645) to Patrick Geddes, who applied it, at the beginning of the twentieth century, to the urban agglomeration provoked by industrialization based on the use of coal and that "not infrequently propagates itself as a mass of relatively uniform density over (...) hundreds of square kilometers" (p. 559). Mumford extends the use of the term to apply it to the characteristic process of the twentieth century by which little separated villages and the suburbs of great cities become aggregated in an undifferentiated and shapeless urban mass in which there are no centers or an active urban life (cf. p. 632). He maintains that this generalized growing density and the propagation of the urban mass are no longer produced by "the original forces that created conurbation", the coal and the steam machine, but they "were supplanted by the net of electrical energy, by the electrified railroad, and latter by the automobile and highways" (p. 645); and that conurbation is the main force that "altered the overall equilibrium between urban population and agricultural population" (p. 632).

Mechanization and urbanization of the field, however, constitutes only one aspect of the process of consolidating the technical environment - the establishment of a technical stage that would be maintained simultaneously with later kinds of development of the agricultural technical environment introduced from the 1960's on. The consolidation of this stage evidently occurred at the same time as the corresponding transformation of the agricultural activity of producing food and also, significantly, beginning in the second half of nineteenth century, of the discipline of Agricultural Science into a specialized area of engineering, whose earlier phases remained closely connected with working in the fields, but with later ones, following the advent of genetic engineering, important parts of traditional agronomical activity, such as the selection of seeds, would be transformed into laboratory and experimental specialties.

Meanwhile, two others stages would be added to this first stage of the transformation of agriculture into an industrial activity. (They correspond to the deepening of industrialization by means of further mechanization and automation).[2] One, commonly called the "green revolution", involves the introduction of industrial products of chemical engineering, particularly, chemical agro-toxics and fertilizers, used intensively in agricultural production with the aim of increasing economic returns rapidly. The other is that of intensive transgenic agriculture, following the development of biotechnologies informed by genetics, and molecular and synthetic biology, in particular the development of transgenic techniques which make possible the production of GMOs, conceived as biological machines. (Chemical engineering of the second stage is deepened by genetical engineering of the third; the development of specialization in Agricultural Science shifts from the field to the laboratory, from agronomy to genetics, and in the direction of experimental decontextualizing strategies).

Footnote

[2]For a preliminary discussion on the repercussion of automation processes in the world of work, see Holzmann (2011), Ellul (1962), Merton (2013 [1947]).

Friedmann was preoccupied with a more encompassing characterization of technical environment that cannot be reduced to the impact of transformations in agriculture, and hence in the natural environment. He takes the "technical environment" to refer to "that environment developed in industrialized societies and communities since the beginning of the epoch of industrial revolutions" - i.e., from the end of eighteen century and first half of nineteenth century - in which "a network of complex techniques tending towards automatism comes more and more to entangle human beings" (Friedmann, 1968, p. 77). They include: productive techniques in industrial and in agricultural sectors; techniques of administration and distribution; techniques of consumption, including those that tend to transform domestic and family life; the various techniques of transport; and the techniques of communication. In addition, the "technical environment" brought with it a type of "psycho-sociological conditioning" (p. 81), in the sense that this environment provides or facilitates a certain type of psycho-social relation that tends to weaken social inhibitions towards the danger of accelerated environmental transformations provoked by technological applications.

It is important to keep in mind - especially in the discussion of the intensive introduction of biotechnologies in agriculture - that the distinction between natural and technical environments is neither absolute nor strict, since it makes no sense to treat human beings separately from their dialectical (mutually reinforcing) relations with their environments; thus, the expression "natural milieu" refers to an environment that "since pre-historical times [has been] a relatively technical environment" (Friedmann, 1968, p. 33; cf. Martins, 2012, p. 91-2). While domesticated plants and animals, for example, can be considered as artificial to a degree, modified by the "technical" process of domestication, they are clearly outcomes of the biological co-evolution of organisms and environment, just as human beings are. But clearly, the extended modification caused by technoscientific (anthropic) transformations of the environment trigger natural processes that fall well outside of the domains in which the technical capacity of control can be exercised, for example, the acceleration of climate changes that are threatening the survival of the human species.

Finally, this presentation of Friedmann's ideas may be summarized using the incisive formulation of Jacques Ellul: "the technological order has become the new and specific environment in which human beings are required to exist, one which has supplanted the old one, that of nature" (Ellul, 1962, p. 394). I do not subscribe Ellul's thesis of the complete replacement of the natural by the technological environment, although I recognize that the natural environment in which we live is highly mediated by technology and technical devices.[3] Consequently to say that our (post-modern) societies are technoscientific societies means that science and technology are subordinated to an economic order that evaluates and organizes them in terms of the likelihood of their contributing towards high productivity, profit and capital accumulation, or towards economic growth. Here, it is important to keep in mind that, while in current technoscientific societies it is the economic order, that is, the economical sphere of social activity, that serves as the vector that orders the technological system, in such a manner that different technical bodies (institutions) or subsystems gain priority according to their economical function (profit, capital concentration, market expansion etc.), in other civilizations there are other kinds of ordering vectors. In Ancient Egypt, for example, the vector is a transcendent metaphysical order, that of the possibility of an eternal life for which this earthly life is taken as a preparation, in such a way that it is the theological religious sphere that directs the hierarchy of technical systems designed to constitute the technological device that guarantees the passage from this earthly life to eternal life. The important concept which can be found in Ellul's conception is that of "technological order".[4]

Footnote

[3]To say that we live in a technoscientific society does not imply that nature has been completely supplanted or that the natural environment has disappeared, although natural tendencies are disappearing and being substituted or rivaled by artificial tendencies of the social technoscientific environment. On the contrary, the acceleration of climate changes, hurricanes, massive floods, prolonged droughts, and other natural catastrophes, to which human beings must adapt socially and organically, may be considered to be compensations of nature in response to the technical modifications imposed on the environment.

[4]In general, Ellul uses the term "technique" in a wide sense that involves "not only the ways in which one influences things, but also the ways one influences persons" (Ellul, 1962, p. 394, note *), so that "technique" can be considered as the near equivalent of "technological order". But in some contexts, Ellul seems to use the term to designate particular procedures or technical solutions. In this article, I will use "technology" to refer to the system of techniques, this is, to the set of techniques and technical normalizations or regulations, and I will use the term "technique" when there is reference to uses and applications of particular techniques. Thus, genetic engineering is a particular technique pertaining to biotechnology, that is, to the set of molecular and genetically-based techniques.

SOME FEATURES OF TECHNOLOGY LINKED WITH RISK APPRAISAL

Before analyzing the impacts and risks that follow technoscientific transformations of agriculture and the highly artificial environment produced by it, in this section I will point to some features of technology that are especially important for the ethical dimension of technological appraisal, as indicated by Jacques Ellul (cf. 1962, p. 412-21) in the context of analyzing what can be called the "myth of technological progress". Ellul defends the thesis of the ambiguity of technological progress: "it cannot be maintained that technological progress is in itself either good or bad" (Ellul, 1962, p. 412), based on an argument in which he maintains that "contradictory elements are always indissolubly connected" (p. 412) to the development of technology. The thesis is manifested in four main features of the development of technology:

(1) All technological progress exacts a price;
(2) technique raises more problems that it solves;
(3) pernicious effects are inseparable from favorable effects; and
(4) every technique implies unforeseeable effects (Ellul, 1962, p. 412).

I will now analyze each of these features from the perspective of its role in evaluating technological applications in agriculture.

Openness and Closedness of Technological Applications

First, "all technological progress exacts a price" does not so much refer to a price in money or in intellectual effort, as signify that "when technological progress adds something with one hand, it inevitably subtracts something with the other" (p. 412-3). Ellul links this contradictory element to the fact that, "viewed objectively, technological progress produces values of indubitable merit while simultaneously destroying values no less important. Consequently, it cannot be maintained that there is absolute progress or absolute regress" (p. 413).

I will adapt Ellul's argument by saying that the development of technology is submitted to a kind of dialectics between openness and closedness of possibilities in technoscientific applications,[5] that is, between the new possibilities opened by new technological development, and those that at the same time are closed - so that in extensive technological applications generally, accompanying the realization of new possibilities, there is a reduction of possible courses of actions and consequently of choice.[6] This means that decisions about implementing a technical innovation should take into account (or be guided by) the dialectics between openness and closedness in a context that makes comparison with possible alternative applications. Only by subjecting each technological project to comparative analysis or examination can one identify clearly the novel possibilities that may be realized and those that may be destroyed.

Footnote

[5]My argument is freely adapted from (without pretend to be an exposition of) the anthropologist Gérard Chazal (2007, p. 67-8). It is also related with what Feenberg (2009, p. 75-80) calls the "intertwined hierarchy of technology and society", in which there is a process of reciprocal construction between technology and society. Truly the two aspects - dialectics of closedness and openness of possibilities and intertwined hierarchy of technology and society - are both cases of an universal invariance which consists in the final analysis in the interactive relation between organisms and environment or, to put it anthropologically, in the relation of humans and the environment.

[6]In certain sense the most powerful instances of the effectiveness of human will in the world are also the ones that narrows the space for the exercise of human will - a paradox that puts into question the limits of technical control of the world.

Consider, for instance, transgenic agriculture, clearly a product of technoscience. It represents the implementation at large industrial scale of very specific technological possibilities such as the application of transgenic techniques aiming to produce seeds resistant, for example, to certain chemical herbicides. Hence, GMOs enrich and expand the possibilities opened up by the Green Revolution that led, in the 1960s, to the chemical industrialization of agriculture with the large scale use of agrotoxics and chemical fertilizers, and to the increasing artificiality of technical environment in the fields and generally in agriculture. However, in virtue of the intensified use of agrotoxics in large plantations of GMOs, soil and water pollution increased and accelerated, insect and amphibian diversity, as well as the variety of seeds of the species under cultivation, were reduced in function of contamination by transgenic genes, and reduction of native species. These consequences add up to a dramatic reduction of natural assets (which become ever more scarce with the advance of technical environment). They also undermine the possibility of developing alternative technologies, such as agroecology, that explore less destructive and more harmonious possibilities, and that are more dependent on natural environmental assets, for example, the existence of fertile, non-depleted, unpolluted, recoverable soils.

Technological Problems and the Role of Alternatives in their Solution

The second feature is that "technique raises more problems than it solves" (Ellul, 1962, p. 412, 414-7). It involves that technology develops "by solving a certain number of problems and by raising others "(p. 414, which cannot be resolved by the same technique that gave origin to them, or that"there is never *one* technique, which solves*one* problem" (p. 415, italics in the original). This means that applying technology to solve problems is, on the one hand, a systematic issue that depends upon the set (or network) of techniques that compose the whole system or the "technological order" and that can explore the technical possibilities made available by the system. On the other hand, a technological solution is always a technical solution to a single problem considered by itself, so that the solution is decontextualized, partial, and very specialized, and when used at large scale in the technical environment produces a series of effects at various of its levels that require other techniques to be dealt with. This ambiguity clearly points once again to the necessity of taking into account the range of possible alternative techniques permitted by the technological system of a certain epoch.

All this can be seen as part of "the vast proliferation and diffusion of the technological way of life, diffusion that is as much vertical and intensive in what is called 'colonization of the living world' (Lebenswelt), as horizontal and extensive on the planetary scale: the technological monoculture" (Martins, 2012, p. 160). Thus, the predominance of GMOs monocultures in the domain of agriculture is not surprising, it being an exemplary case of technological colonization of the food system, that through extensive and accelerated technical transformation of the environment, brings it about that unpredictable and undesirable effects of the generalized application of highly specialized techniques become consummated facts. These consummated facts are then used to put constraints onex *post facto* regulation, which again is always a technique, in virtue of the technological environment in which they are inserted, and which consolidates in the social order often irreversible consequences - including irreparable losses, for example, connected with plant diversity and soil fertility - for the stock of natural assets. There is thus a constant societal process of assimilation of the undesirable consequences by means of which society adapts, through technical and juridical regulation, to the environmental transformations so that efforts to introduce social and ethical restrictions of undesirable or even bad technological consequences are inhibited; this serves as a technological device to maintain the process of "colonization of the living world".

Evidently the above argument contains the tacit supposition that "only technology can resolve the problems occasioned by technology", because every technical solution is introduced into the technological order or system, which is a network of social spheres (economic, educational, law), of areas (urban, sanitary, environmental), of levels (local, regional, global), of scientific disciplines, and of technical specialties derived from experimental knowledge that these disciplines make available. The technological system is therefore an open system that makes available a broad set of particular techniques applicable in the areas, and on which the successful applicability of techniques depends on the natural and artificial assets of the present configuration of technological environment. In this sense, even an alternative solution, which is environmentally favorable and economically sustainable, is still a technical solution, and cannot be otherwise. However, although the technological system is an open system, it does not escape being subject to the economic order; that is, the various technical possibilities are arranged in an hierarchy and favored depending on their contribution to economic development and growth.

But to accept the supposition that there are only technical solutions to problems generated by technology does not oblige one to adhere to technological determinism, that is, that there is only one type of technique available that could deal with the problems generated by particular technical solutions, or in other words, by technical solutions for problems defined inside artificially restricted contexts, that are experimentally controllable (hence, rigorously decontextualized), and so produced within closed technical sets of solutions. A technological version of the argument for the best possible world is insinuated here, one that Martins calls ironically the "theorem of Panglossian existence": "there are always technological solutions for the problems created by technologies (especially biomedical and ecological problems)" (Martins, 2012, p. 161). However, to consider this proposition as a theorem of technological determinism depends on the degree of assent given to the suppositions:

(1) that the solution presented is the best possible solution, in fact, the only one which is appropriate (and efficient);

(2) that "there are no alternative technologies that are appropriate, intermediate, and with more favorable outcomes" (Martins, 2012, p. 160).

Both suppositions face great difficulties. Regarding the first, technical solutions are inevitably incomplete, since they are solutions for problems defined within artificially restricted (rigorously decontextualized) contexts. When they are applied on a large scale in the highly artificial environment of the technological order, they engender an inevitable "residue of new social technical problems due to the incompleteness connected with interactions and limitations" of the closed technical sets and of "their unpredictable secondary effects" (cf. Martins, 2012, p. 162). The latter are the effects of the large scale causal propagation through highly artificial technological environments of specialized technical solutions, that is, solutions whose technical efficacy was narrowly considered within the boundaries of the closed technical set to which they pertain, and so they generate other technical problems that have to be treated by other technical sets:

(...) this thesis of the inextricable and unceasing chaining of problems, solutions, and new technological problems (...) [implies that] new technological or technical-social problems proliferate more rapidly than solutions, and that the set of residual problems become more difficult than previous ones due to factors such as the dynamic of technology, growing complexity [of technological system], increasing costs and decreasing resources, exigencies for greater control, and the inertia of social institutions (Martins, 2012, p. 162).

Furthermore, the tacit presupposition that there are no alternatives expresses only a determinist understanding of "technical efficacy". This can be questioned from the perspective of the difficulties related to the first supposition and also because it reflects the hegemony granted in scientific institutions to decontextualizing (experimental and reductionist) strategies. Hence, it favors technical solutions based on experimental research of controlled causal mechanisms in the restricted environment of the laboratory because these solutions involve the increase of control of specific artificial and natural processes of closed technical sets that compose our technological environment. Decontextualizing strategies are not the only scientific strategies that may orient efficient technical applications, however; and they are incapable of solving the residual problems that propagate through other spheres of the highly artificial technological environment.

The solution of residual (technical/artificial and natural) problems, that are produced by negative effects or unexpected accidental consequences of specialized technical solutions, can only be treated by open technical sets that make use of contextsensitive strategies or even of a plurality of strategies, for appraising the efficacy of a solution requires taking into account the full range of both intended and residual effects of a technical application. Then, when making decisions about applying large scale technical solutions, it becomes possible to consider alternative technical solutions that balance economic with social and environmental efficacy.

But while the technological version of the Panglosian theorem of existence secretes technological determinism, the market version of this theorem secretes economic determinism. Following the neoliberal version of Panglosian theorem of existence:

Every problem caused by markets can be solved in a timely manner by more and better markets (whether these problems are economic, social or ecological), and not by non market behavioral patterns and models" (Martins, 2012, p. 163).

The argument for economic determinism depends on the same kind of suppositions that are used in the argument for technological determinism, and whose difficulties I have just presented. Regardless of its value as a determinist principle, however, the Panglosian theorem represents an attempt to close the cycle - technological and economic - of the efficacy of technical solutions, excluding the possibility of alternative technical solutions. Thus, in the economic version, the negative residual effects of a technical application are considered "negative technological externalities or non commodities (...), the environmental disasters such as the famous 'tragedy of the commons' can be solved by market solutions and only by market solutions" (Martins, 2012, p. 163). This means that residual effects of technical-chemical solutions in agriculture, such as air and water pollution, should receive a technical solution compatible with the economic efficacy, which limits technical efficacy to the closed agro-biochemical technical set that is exactly the technical set that cannot be developed without causing air, water or food pollution. These latter remain as residual problems without an economically viable technical solution. Then, these problems tend to receive a market solution: either insurance that pays monetary indemnities to those affected by the irreversible consequences caused by the solutions of agro-bio-chemical techniques, or transforming the question of pollution into an economic market, as in the case of carbon credits.

Again, all this is well illustrated by the agricultural technoscientific solution introduced to resolve the technological problem of producing food in chemical polluted areas, by means of the accelerated artificial production of seeds and the expansion of intensive monoculture, that generates a chain of effects such as increase of chemical pollution in the environment, reduction of the variety of cultivars, decrease of insect populations (or mutations that generate "super-insects"), disappearance of pollinators (bees and others), degrading of ever more extensive areas of arable soil (and the consequent necessity of expanding the frontiers of farmlands and the probable destruction of the Amazon Forest), creation of conditions favorable to the appearance of "superweeds" etc. Furthermore, transgenic agriculture contributes to put pressure of society's capacity to deal with health problems such as the increase of breathing problems in children and aged people, the continuing process of displacing people from rural to urban zones - problems that can only be solved by other techniques available in the technological order, or that are linked to another conception of technical efficacy.

The Inseparability of Beneficial and Harmful Effects of Technology

The third feature concerns the effects and consequences of applying technologies to the solution of problems. It can be expressed by "the thesis of the inseparability of beneficial and harmful effects of using technologies", which has an immediate consequence at the ethical level. This thesis rejects outright the idea, which is basic and central to much common moral thinking, that technology is intrinsically neutral, and that whether it is used for beneficial or harmful ends depends on decisions made by its users. This idea generates (in the light of the inseparability thesis) the illusion that there need not be harmful consequences or that they can be cancelled by "good uses". Consider the example of nuclear technology. Whether it is used for peaceful or military ends, it is implicated in harm (for example, nuclear waste that must be kept under constant surveillance since it is highly contaminating for a long period) that cannot be ignored on the ground that it is intrinsically connected with the beneficial effects of the uses of nuclear technology; it is impossible to generate energy for the good of people without generating very harmful toxic waste. Another equally obvious example is that of the use of genetically modified seeds, which cannot be used without causing loss of diversity (variety) of the crop species planted.

This feature is particularly relevant to decision making about possibly implementing dangerously polluting chemical technologies. Moreover, the thesis of the inseparability of beneficial and harmful effects can lead to critical questioning of what is commonly called "standard risk analysis". It leads to recognizing that both the terminology employed - "risk" instead of "harmful consequences" -, and the consequent classification of risks as "potential" or "real" or "effective", involve the presupposition that technology is neutral, for it produces manifest benefits, and only potentially occasions risks (and harmful consequences). This presupposition is manifest once again when evidently harmful effects such as the deaths of many people (as in the case of the leaking of gas in Bopal chemical industries) or the demonstrated loss of environmental diversity are called "real risks", implying the supposed intrinsically beneficial character of the technology, relegating the harmful consequences to the level of error or bad use. This takes issue with the utilitarian (clearly economic) interpretation of the concept of "risk" promoted by the large corporations, and the implicit reduction of all value to economic value (profit or loss). Standard risk analysis is fundamentally an organizational or management (technical) tool that serves for calculating the probability of economical loss in connection with the application of a particular technical innovation, and of the corresponding costs of unpredicted residual effects of the large scale use of products developed under experimental controlled conditions. In fact, it is part of a technological device of power that, as we shall see in section 3, is exercised so as to favor (make possible), always in the name of economical progress or development, social reluctance to address the negative consequences of technology (cf. Fressoz, 2012; Agamben, 2007).

Unpredictability and Uncertainty of Technological Applications

The fourth feature of the ambiguity of technological progress is expressed in an universal mode by Ellul: "all technological progress contains unforeseeable effects" (1962, p. 419). When new technical means and new technological solutions are discovered, they are applicable to precisely delimited technical spheres (facts) in such a way that certain results (effects) are expected and obtained. However, in virtue of the highly artificial environmental and the (open) interactions of the expected effect with the totality of effects of the system "there are always secondary effects that were not anticipated, and that in the first stage of the technical progress in question could not in principle have been anticipated" (p. 419-20). This unpredictability could be avoided only if the complete experimental exploration of all possibilities permitted by the technological system at every moment was possible, and obviously it is not.

Note, in the first place, that this points to a strong link between predictability and experimental control. Unpredictability means that we are dealing with variables that were not taken into account in the experimental design which leads to a particular technical development. In second place, since the aim of technology is to control technically the events that occur in the world, complete predictability, that is, the possibility of conducting experiments concerning all objects and dimensions of the technological system (the "complete possibility of experimenting in every sphere") remains as an ideal, which maintains, on the one hand, relations of mutual reinforcement with the value of control, but imposes, on the other hand, the recognition of the limits (boundaries), sometimes very narrow, to which technical solutions are subjected.

Ellul illustrates the thesis of technological unpredictability by means of examples of unpredictable effects in three technological spheres: pharmaceuticals and chemistry; urbanism; agriculture. (This will give us the opportunity of returning to the reflection on the transformations of natural into a technical environment.)

Concerning pharmaceutical and chemical technologies, Ellul discusses three cases: aspirin, DDT and thalidomide. Aspirin has remarkable effects connected with eliminating headaches, but its prolonged ingestion has the undesirable collateral effect of increasing susceptibility to severe hemorrhages. DDT, the first agrotoxic used extensively and intensively, was considered in 1945 to be "a prodigiously successful means for the destruction of all kinds of vermin and insects" (Ellul, 1962, p. 420) and absolutely innocuous to human beings. But, when consequently it was intensively applied on the surface of the planet, it turned out to have the unpredicted secondary effect of causing anemia in animals and in humans. Regarding thalidomide, Ellul just mentions its terrible genetic and teratological effects, and reiterates that "identical problems [that is, of producing harmful collateral effects] are potentially raised by all chemicals consumed by animals or men" (p. 420). The upshot of this discussion is the following characterization of secondary (collateral) effect: secondary [collateral] effects: effects which are essentially unpredictable and only revealed after the technique in question has been applied on a grand scale (...) (Ellul, 1962, p. 420).

This characterization of collateral effects is relevant for the exercise of the Precautionary Principle and points again to the necessity for more scientific research on the collateral effects of the continued use of drugs and of toxic substances on animal and human health, in order to anticipate those effects that are unpredictable in the short term. (Obviously, the health of plants, animals, and humans are linked by the food chain, and polluted air and water undermine environmental health.)

Another kind of technological unpredictability is found in the urban dimension of the technical environment, more directly linked with psycho-sociological effects on the behavior of the inhabitants of large cities, because in the urban planning we are very specially "confronted with the effect of the technological environment on human beings" (Ellul, 1962 p. 420). In large cities, feelings and sensations of isolation, loneliness, absence of human contacts, physical and mental stress, fear of violence, etc, are common. Large scale urban projects, like those of Le Corbusier, that were based on the diagnosis that these problems originate in the fact that "big cities dwellers do not know one another", and that, in order to favor human contact and redirect the psycho-sociological effects, involved the construction of "large blocks of dwellings where people could meet one another as they would in small cities, with everything (grocer, baker, butcher) included in the block so that people would get to know one another and a community would come into being (...)" (Le Corbusier *apud* Ellul, 1962, p. 420). However, the psycho-sociological result of constructing the large housing blocks, instead of being a return to an idealized community life inspired in villages, was the opposite of that predicted, for the problems of solitude and isolation became aggravated by the significant increase of violence that erupted in these urban agglomerations. In fact the construction of large housing and residential blocks ended up being integrated into a technological device aiming to solve two basic urban problems: housing and locomotion. Many programs of slum ("favela"; "villa-miseria") removal in Latin America followed this model of construction of big agglomerates of housing blocks. This solution also produced part of the process of suburbanization on the peripheries of large cities, and the destruction of natural environments as well as the reduction of productive farmlands (cf. Mumford, 2008 [1961], cap. 16-17).

The other part of the process of urbanizing rural areas was a consequence of the overwhelming predominance of automobile; and, in American countries, the predominance of transport moved by burning fossils fuels led to constructing a network that inter-linked cities based on roads and freeways that cut through rural zones, many of which are dedicated to food production. It led also to an urban network of avenues, tunnels, bridges, viaducts and parking lots that clearly prioritizes the individual automobile (cf. Mumford, 2008 [1961], cap. 17). This huge technological system of transport enabled the urbanization of the rural areas increasingly distant from the suburbs of large cities by means of developing urban projects - promoted by the construction business and directed towards the middle class - that create gated communities with controls on the entry and exit of people, and that often transform farms and fertile and arable lands into urban condominiums (cf. Mumford, 2008 [1961], cap. 16). From the psycho-social point of view, this kind of model of community, far from allowing a life closer to nature, supposedly calmer, more balanced, and favorable to rest and leisure, ends up producing unexpected difficulties, and becoming a source of physical and mental stress. On the one hand, people living in these artificially created communities tend to generate a psychological environment of suspicion and censorship, in which "people are perpetually under the eyes and the surveillance of their neighbors" (Ellul, 1962, p. 421). On the other hand, the growth of rural urbanization leads rapidly to traffic jams at the entrance points of the great cities, requiring more and more time to travel between one's distant residence and work place in the city. This example shows how urban planning solutions, broadly technical ones from the economic, managerial and engineering point of view, can produce unexpected and even psycho-social consequences that are contrary to those expected.

Intensive conventional agriculture, "the massive cultivation of certain plants like corn and cotton" (p. 421), practiced in the United States, Brazil, Russia, and Argentina among other countries, provides Ellul's third example of unpredicted effects of large scale technological applications. From the technological point of view, the extensive plantations (monocultures) of corn and cotton have represented economic progress because they produced an enormous increase in productivity. Deforestation also resulted in economic progress, doubly profitable because of commercialization of woods from the destroyed forest and opening up new areas for cultivation and hence productivity. However, it was not anticipated (predicted)

that corn and cotton are plants which not only impoverish the soil, but even annihilate it by the twofold action of removing certain natural elements and destroying the relation between humus and the soil particles. Both are destroyed by the roots of cotton and corn to the extent that, after 30 or 40 years of cultivation of these agricultural products, the soil is transformed into a veritable dust bowl. A strong wind need only to pass over it to reduce it to bare rock (Ellul, 1962, p. 421).

Intensive agriculture has similarities with the first example of pharmaceutical and chemical products. Once again there is an unpredictable (in the short term) collateral effect that becomes manifest only when the technological solution is applied at large scale over a long period of time, in this case, thirty or forty years. What the second and third examples have in common is that they are part of the same process of technical transformation of the rural environment that results in the continuing decrease of fertile areas that could be used for production of food at small scale in the peripheral areas of large cities. In addition, in the huge productive areas of the interiors of American countries, the disaster of green revolution, which in part is pertinent to Ellul's examples of corn and cotton, is transformed by transgenic agriculture into a catastrophe for the environment and health. The large scale applications of chemical, genetic and biochemical solutions greatly increase the negative collateral effects on the environment, what the economists call "environmental externalities", obliging an increase of social tolerance to risks and harmful consequences. There is an acceptation that the environmental externalities of large scale technological applications, such as reduced water resources, chemical pollution and soil infertility, the residual cumulative effects of continuous ingestion of GMOs etc. are the price that we have to pay for progress, modernization, development and, more recently, innovation.

There is a powerful compensatory mechanism (that removes social inhibitions) that generates continuous adaptation between the transformations of the technological environment and the social practices that unfold in it. This mechanism is a technical device that has the components: standard risk analysis; insurance against damage and risks, and regulation of techniques, that is, the introduction of technical codes and legislation about the limits of use and application of certain techniques. The function of this procedure is to engender acceptance that, although the techniques may have dangerous or harmful effects, their benefits justify having tolerance to the risks they occasion.[7]

Footnote

[7]There is no space here to deal with the economic consequences of technological unpredictability and, in particular, of the economical aversion to uncertainty (cf. Martins, 2012, cap. 5, p. 180-8).

TECHNOLOGICAL RISKS AND THE STATE OF TECHNOLOGICAL EXCEPTION

There is a precise anthropological sense in which the complementary relation between humans and the environment in which they live is what defines humanity. The human is its environment; and the environment in which humans beings live is the measure of human intervention in nature; and the degree of artificiality of the environment is in the final analysis the measurable expression of our separation from nature. The further the transformation of the natural into a technical and artificial environment proceeds, the more we subject ourselves to the unpredictable (in the short term) consequences of technology and to the uncertainty of our partial technical solutions, and, as in the case of agriculture where biotechnological and chemical solutions are applied at large scale, we run the risk of environmental and health catastrophes and losing natural assets to such an extent that we even endanger the survival of the human species. Technoscientific agriculture confronts us with "the violence of danger that eliminates all protective zones and all modern differentiations" (Beck, 2010 [1986], p. 7). I will deal with this notion of "risk" that is used to characterize society in the present state of development of technology.

In a simple way, "risk" is synonymous to "danger", in other words, risk expresses the possibility of danger, that is, it expresses the high probability of the occurrence of an inconvenient effect or a fatality, whose occurrence cannot be controlled by the same particular technical solution which make possible that particular technical action or application. Risk also applies to the unpredictable (in the short term) consequences and to the residual effects (produced by the effect of causal propagation through the technological environment) of technical solutions applied at large scale.

"Risk" is an anthropocentric notion, from which follows its ethical dimension, for it would not be risk without human implication, without some kind of technical action (cf. Bourg, 2007, p. 123) on the environment, without having repercussion on what is to be an human in such an environment. This is why in the dimension of human social organization the notion of "risk" acquires an important juridical sense, according to which in insurance contracts, risk is understood as an incident (an economic loss, a physical or material harm) which demands indemnity. Risk, in the juridical sense, refers then to the responsibility or the duty related to the loss or harm caused by a technical action that produces such a risk situation. In the economic sense, however, a risk is evaluated by the economic (cost/benefit) standard risk analysis, and it can be defined as the association between an eventuality (probability of economic profit in an technological endeavor or application) and a vulnerability (probability of harm to people, individual goods or society; harms that evidently will result in losses to be covered be insurance or paid by monetary indemnities) (cf. Bourg, 2007; Dusek, 2009, p. 89-97).

The philology of the word "risk" is clarifying. The term appears in the West in the middle of twelfth century, when Pisa and Genoese merchants borrowed from "arabic the word 'rizq' ('the part that God attributes to each men') to designate in their contacts the losses and profits linked to uncertain events" (Fressoz, 2012, p. 27), that is, to increase the profits and diminish the losses (or insured part of these) in function of danger. Throughout the fifteenth century the term "risk" was in regular use among Portuguese and Spanish navigators in connection with financing the technical and economic adventure represented by great feats of navigation and by the colonization of America. This means that the insurance system linked to economic risk, as well as the accounting system registering receipts and payments, were part of mercantile capitalism, long before risk was incorporated in the insurance system of industrial capitalism during the chemical revolution of eighteenth century and its technical application on a large scale during the first industrial revolution after the French Revolution.

Maybe it was Jean-Jacques Rousseau who first used the term in the sense of the expression "technological risk" referring not only to the economic sense of insurance, but applying it to the urban planning (technical) conception of the cities. Differing from Voltaire's revolt against the "violence of nature" in the earthquake that destroyed Lisbon in 1755, Rousseau maintained that the responsibility was in greater part human, attributing the catastrophe to the risk to which inhabitants of Lisbon had exposed themselves for "having grouped twenty thousand dwellings of six or seven floors" (Rousseau apud Bourg, 2007, p. 123-4); and many people had died looking for their belongings, an attitude that reveals the ignorance of the danger to which they were exposed.

During the historical process of constitution of the technological system - which is a fundamental part of the process of scientific-technological rationalization of modern societies that has continued from the mid eighteenth century until today - there was consolidated a social perception of risk, molded in a technical-economic conception of the same risk. This conception is based in the standard risk analysis, which is an instrument of the economic rationalization of profits and losses being part of a technological device - in major part mathematical, constituted by probability, statistics, game theory and rational decision theory - that enables the direct technological dangers and its eventual negative social and environmental consequences to be absorbed by the market. This is done by the financial and actuarial processes of insurance that have accompanied industrialization since the first Industrial Revolution until the chemical and biotechnological industrialization of present agribusiness.

If we look at the great technological interventions that marked the transformation of the natural environment into the present technological order, an environment more populated by technical objects and even artificial living beings, we can detect the mechanism by which the negative consequences and residual effects are assimilated, and come to be considered as acceptable (natural, inevitable) integral parts of the environment. This technological device makes viable a process of social adaptation to technology and to the risks, harms to health and environment that it occasions, that are incorporated by means of what Fressoz calls "the ontological form of the modern neutralization of inhibition", taken in the sense of "the set of descriptions that aim to adjust the world to the technological imperative" (Fressoz, 2012, p. 286) of modernization and economic development. And, in fact, the neutralization of social inhibition seems to be more effective when it is established "in the immediacy of the relation between the world and the techniques" (p. 286), because in this manner it obtains the aim to "weaken the technological impact and neutralize the critical sense of the accidents, in order to capture, orient and align the perceptions and the behaviors in the sense of technique" (p. 287). There is then a constant process of social neutralizing of inhibition by which risks and accidents, intrinsically linked with technical solutions and applications, are incorporated as parts of certain technical sets of the technological order, and in this manner after a long period of regulation - which can never be completed - become normal, controllable within certain parameters, and so are neutralized in the technological order; at the same time social inhibitions concerning risks are weakened in view of modernization, technological progress and economic development.

Furthermore, the process of gradual construction of the technological system that is composed by the network of technical devices embodies a historical systemic invariant, namely, the exceptionality of the social, health and environmental situation created by every new large scale application of technical solutions - a state of technological exception (that in itself does not require explanation, and that is tacitly taken for granted), in which values and "the traditional norms that rule health, property, environments, imputations of harms, and responsibilities" (Fressoz, 2012, p. 288) are suspended. In this exceptional state, certain matters concerning artificial beings, new technical devices of control, and the inevitability of residual or even bad effects of technological solutions are able to become consummated facts. While the state of technological exception endures, incessantly more and more of such consummated facts about environmental, social and health risks are created, and they have impact in juridical debates and *ex post factum* technical regulation of its limits. The state of technological exception corresponds, then, very significantly to the maintenance of a state of suspension of rules, laws and established order, to a suspension of social and ethical values and particularly, as we shall see in section 4, to suspension of the values of scientific ethos, thus to adopting an antiscientific posture. It authorizes the execution of an open-ended process of experimentation that always involves increasing the technical component of the environment, and that produces effects and consequences which are very far from being predicted, controlled or avoided.

The juridical manner of solving the state of technological exception consists in closing the technological cycle by means of regulations on (for example) the new artificial living beings, and on the procedures, dangers, accidents or harms caused by them. Beginning with recognizing the consummated facts generated by technology, technical and juridical regulation "redefines the stuff of the world, the beings that compose it, and its regularities, redefining equally the technical forms" (p. 290). However, recognizing a consummated technical fact imposes on science the task of describing and explaining it with the intent of absorbing the exception, setting its limits, and neutralizing its effects in the technological order. Therefore, science and also technique give rise to a procedure of expertise - that supposes a specialized knowledge not only about the disputed matter but also in the process of regulation - by which technology ends up imposing a technical regulation on society, instead of society establishing the social, sanitary and environmental limits of technology. Observe that the process of regulation occurs for the most part in a state of exception and, consequently, of suspension of habitual normative practices, and that it concludes with the normalization of the exception (cf. Fressoz, 2012, p. 289). So, the closed cycle of technical and juridical regulations - which goes from the state of technological exception, in which are generated the technological consummated facts, to the conclusion of technical and juridical regulation - gradually prepares the process of weakening social inhibitions concerning risks, that is, the adjustment of society to the exception by turning it into the "normal", based on the imperative of technological progress and economical development.

This cycle - state-of-technological-exception/consummated-facts/ regulation - can be seen as an invariant component of the historical development that brought about the gradual implementation, from the 1750s on, of the technological order, the introduction of the scientific mentality in the productive system, and gradually in more sectors of society. Modernization and technological progress - that correspond to the advance of scientific rationality and of the bureaucratic order in society - is found in the hygienist movement of prevention of smallpox that agitated France, England and the United States from the middle of eighteenth century, and that is undoubtedly the first large scale application (in public health) of a medical technique derived from the medical study of infectious deceases. The movement began with the discovery that the inoculation of the pus from the pustule of smallpox could make a person immune to the disease; and led, in the course of the campaign to combat smallpox, following an internal technical development, to the discovery of the vaccine and to the technical development of its production, conservation and use in inoculation. In this very large scale experiment with human populations, conducted sometimes far away from controllable conditions, epidemics broke out and, during a long period of time until the stabilization of the vaccination technology, the smallpox inoculation produced many death and, among those immunized, many collateral effects, such as blindness and (most commonly) deformations in the inoculated limb. This huge experiment conducted on the civil population generated revolts and resistance that lasted until the technique was stabilized, its efficacy was proven, and it was regulated as an acceptable (even desirable) practice of public health that could be imposed on the population by the force of the law. In this case, one can see a clear neutralization of inhibitions in the face of dangers and harm occasioned by these sanitary campaigns, in the name of the modernization of society and medical and health related progress (cf. Fressoz, 2012, cap. 1-2).

One can find the same cycle at the very beginning of the first Industrial Revolution in the process of chemical industrialization, with the production of artificial soda financed by the French bourgeois after the Revolution of 1789. Following the discovery of the chemical synthesis of soda and the technique for artificially producing it - due to the extraordinary development of Chemistry during what is called the Chemical Revolution of the eighteenth century - the establishment of factories (of all sizes) that produced soda accelerated, in Paris, Lyon and other French cities and their suburbs. These factories generated environmental disorder in their surroundings, with the air intensely polluted by toxic acid gases that corroded walls of buildings, clothes hanging on clotheslines destroyed, and serious breathing problems caused for those who lived close to the factories, while in the fields there were great losses of crops. The state of technological exception that lasted four decades from 1780 to 1810 was resolved partly with the improvement of the chemical process of soda production that aimed to eliminate the emission of acid gases, and this involved elaborating technical norms of security; and partly by the creation of a system for insuring the industrial capital invested in chemical production, so that the insurance system would be improved parallel to the advance of technical security regulation. Until the end of the state of technological exception there continued to be intense environmental pollution in the cities and fields around the first French industries, and there was intense social and juridical debate about prohibiting this kind of factory and about indemnities for material, environmental and health harms caused by the chemical production of soda (cf. Fressoz, 2012, cap. 3-4).

Finally, the same cycle - state-of-technological-exception/consummated-facts/regulation - was present during the whole nineteenth century with the development of the steam engine and with the transformation it brought about in the means of industrial production (for example in the mechanization and automation of the textile industry) and in the means of terrestrial (trains and railroads), fluvial and naval (steam ships) transport. The whole process of implanting steam technology occurred in a state of technological exception. This had to do, on the one hand, with the accidents produced by steam engines, such as the explosions, the avoidance of which eventually was to depend on the technical development of boilers with internal controls of the pressure of the steam inside of them, and of mechanisms of refrigeration, in order to be avoided; and, on the other hand, with the huge environmental transformation caused by the railroad with the beginning of the production of energy by burning fossil combustibles - first with coal and then coke (cf. Mumford, 2008 [1961], cap. 15; Fressoz, 2012, cap. 5). Here also there was a long process of regulation, maintained partly by the technical process of improving engine security and partly by insurance for accidents and indemnities to victims. The combustion engine, the automobile and the highway transport deepened the environmental consequences of burning fossil combustibles, and extended, with respect to their environmental consequences, the state of technological exception until our own days.

It is not surprising that the process of industrialization of agriculture consolidated in the 1920s with the mechanization of the fields and the advance of monocultures, and continued in the 1960s with the Green Revolution and the advance of the use of chemicals in agriculture and, from the 1990s on, with transgenic agriculture, has occurred in a state of technological exception which has endured for almost a century, and maintained a lasting process of regulation of negative environmental and health consequences of large technological applications in agriculture. The state of technological exception seems to be thus intrinsic to the very idea of technological innovation. In Brazil it serves to neutralize social inhibitions in the face of negative environmental and humanitarian consequences, as in the case of hydroelectric plants being constructed in the north of the country, and of health and environmental consequences, as in the case of agribusiness, always in accordance with the imperative of modernization/development and economic growth. In fact, it permits the exercise of the naked power, in the Hobbesian sense, because it is the power of the stronger exercised regardless of any social pact in a state of affairs in which the usual and accepted values and rules are suspended.

TECHNOLOGICAL EXCEPTION AND ANTI-SCIENCE PRACTICES

The state of technological exception that, as we have seen, is necessary for the development of technical innovation practiced by the large pharmaceutical and chemical corporations permits these corporations to develop a political and rhetorical practice that aims to retard the regulatory process. This practice contributes to prolonging the state of technological exception - and it is antiscientific. Its unscientific character shows itself in three ways: (1) by impeding scientific research on the health and environmental effects of pharmaceutical drugs and chemical substances; (2) by deliberately propagating fraud and ignorance; (3) by continuing to apply and to expand the chemical biotechnological innovations even after scientific verification that they cause harm and after their prohibition in certain countries (mainly in Europe).

The Attack on Scientific Impartiality

The antiscientific attitude is manifested in the fact that pharmaceutical and chemical corporations normally refuse to furnish the data of the tests on the collateral effects of their products carried out in their own laboratories, sometimes even to laboratories connected with the official regulatory agencies. The corporations justify this by alleging that openness about their data would undermine the protections that their patents grant them to maintain secret the constitution of their products (drugs and seeds); and following this allegation, legal prohibitions and restrictions have been put into place on the scientific research that may be conducted on, for example, the collateral, residual and cumulative effects and harms caused by large scale uses of agrotoxics, and on the effects on health of the prolonged and continued ingestion of GMOs (cf. Mariconda & Ramos, 2003).

Impeding access to the information about data related to technical solutions, and prohibiting independent research on its collateral effects obstruct the exercise of science, are incompatible with the scientific method. They prevent the repetition of experiments and comparing data obtained in the corporations' laboratories with those obtained in other laboratories that may involve, for example, variations of the initial conditions of dosage and combinations of chemical substances. These impediments and prohibitions also obstruct experimental research oriented towards the technical improvement of security norms being a part of the regulatory process. Where they apply to research on the collateral, residual and unpredictable effects of specialized technological applications that, as we have seen, generate problems whose solutions can only be found in other scientific disciplines and in other techniques, it is evident that there is external interference - in this case, by economic interests - in determining the objects of scientific research and in applying the scientific method. Thus, we find undermined the autonomy of scientific institutions to investigate, using the standard procedures of the scientific method, the efficacy and the limits of technological applications.

Finally, these impediments and prohibitions introduced in order to maintain secrecy represent a threat to scientific impartiality, the constitutive value of scientific method, the one that stipulates that scientific evaluation should be based only on cognitive criteria (cognitive values, such as empirical adequacy, explanatory power, predictive power, precision, simplicity and comprehensiveness) and the full array of relevant empirical data. The value of impartiality is, thus, intimately linked with procedures for the scientific certification of hypotheses and theories and confirmation of results, based on methods and procedures universally agreed upon that guarantee the objectivity in principle of the knowledge and understanding obtained of phenomena of the world (Lacey & Mariconda, 2012). Because of the secrecy and privacy that they involve, patents stand opposed to the open and public practices needed to ensure scientific impartiality: "the institutional conception of science as part of the public domain is linked with the imperative for communication of findings. Secrecy is the antithesis of this norm; full and open communication its enactment" (Merton, 2013, p. 192; 1957, p. 557). Thus, to impede scientific research based on patent rights amounts to imposing a scientific moratorium, because effectively "the definition of technology as 'private property' in a capitalistic economy" (Merton, 2013, p. 193; 1957, p. 558) is incompatible with the value of communalism, for example, with the conception that the contribution of individual people to scientific knowledge are part of the shared stock (capital) of scientific knowledge. Furthermore, obstructing scientific scrutiny of data and research on the collateral, residual, unpredictable, uncertain effects (risks) etc creates obstacles for developing technical (and also legal) regulations. This antiscientific behavior is aimed at maintaining the state of technological exception.

The Production of Ignorance

The vast agricultural (biotechnological) experiment that is being promoted by agribusiness with the diffusion of monocultures of GMOs - a clear expression of "experimenting in every sphere" (see section 2.4 above) - , generates (unpredictable or undesirable) social, psychological, environmental and ecological externalities (cf. Martins, 2012, p.110). The change of jargon is intentional, in order to better deal with the scientific-technological system insofar as it is structured economically, for agribusiness corporations consider the multiple social, etc consequences of large scale technological applications as externalities, that is, as undesirable effects for which the revenue of crops and its extraordinary contribution to economic development serve as sufficient compensation. In the light of the economic criteria of capital and market, health and environmental consequences and harms are simply the price that society has to pay for technological progress and economic development.

In order to challenge this, since the effects on health of ingesting GMO food products are unpredictable in the short term and, being residual and cumulative, will only be manifested after 30 or 40 years, scientific research needs to be pursued intensely in order to make a more methodical empirical appraisal of the chain of health and environmental effects triggered by this agricultural experiment. In this way, the problems of acting without precaution, and of implementing chemical biotechnology in the shortest possible time, become apparent. The harm caused by industrial antiscientific practices also becomes evident. These antiscientific practices impede the scientific appraisal of the consequences triggered by the technology, and hence risk analysis itself. This appraisal requires that first the residual and cumulative effects that are propagated through the technological order and its highly artificial environment should be investigated; and then the social, health and environmental consequences that follow from these effects and the possible solutions to any problems that they may cause and, at the same time, to evaluate the development of alternative technological solutions to the production of food, such as agroecology, which is potentially highly productive economically.

It is in the context characteristic of the state of technological exception that arises the second antiscientific practice (with serious ethical consequences) of the deliberate production of error, or "fabricated ignorance" (Martins, 2012, p. 73). Ignorance has always stood between justice and utility, and claiming to be ignorant of the consequences of practiced acts has always served as an alibi in courts of justice, because where there is genuine ignorance there is not the ethical intention to perform an evil act. It is important, however, to distinguish two kinds of ignorance.

On the one hand, there is "genuine ignorance", that is, that ignorance produced by technological development (progress) which generates risks and uncertainties that cannot be eliminated, "not only because we are ignorant of the causal mechanism, but because no degree of knowledge would be able to eliminate the uncertainties which follow from the interactions between the new social-technique systems and the biosphere" (Martins, 2012, p. 73). Genuine ignorance is evidently linked to another of the values that are components of the scientific ethos in Merton's sense: organized skepticism, to the demand of detailed critical scrutiny of scientific contributions. This is a consequence of recognizing the conjectural/hypothetical character of science and leads to recognizing the fallibility of scientific knowledge; and effectively skepticism appears as a fundamental characteristic of Popper's conception that scientific hypothesis (theories) should be submitted to severe and rigorous empirical tests in a context of comparison with alternative hypothesis or theories (cf. Popper, 1965, cap. 5). For Merton, organized skepticism is intimately related to the other constitutive values of scientific ethos: universality, disinterest, originality, communalism, and it is conceived as " both a methodological and an institutional mandate" that prescribes "the suspension of judgment until 'the facts are at hand'" and involves "the detached scrutiny of beliefs in terms of empirical and logical criteria" (cf. Merton, 2013, p. 197; 1957, p. 560).

On the other hand, in the last quarter of the twentieth century, "fabricated ignorance" has been spreading in "market economies in consolidated democracies". It consists of campaigns to produce ignorance (...), doubts, uncertainties concerning well founded scientific thesis, with spurious arguments, distortions of results, data from badly conducted research, testimonials presented by specialists in courts to raise doubts about sound results of scientific investigations (Martins, 2012, p. 173, note 18).

The campaigns of production of ignorance have occurred repeatedly from the end of the 1970s - and they have become a common practice of the tobacco industries (Proctor, 1995),[8] petroleum companies and more recently chemical-biotechnological corporations. They aim to induce ignorance by means of generating doubts on themes or results about which there is an extensive scientific consensus.[9] Thus, despite the fact that it had been well established scientifically that there are strong statistical correlations between smoking tobacco and occurrence of lung cancer, the tobacco industry was able to postpone regulation of tobacco by means of an organized program aimed at putting into doubt and discredit scientific evidence about this connection.

Footnote

[8]In this book Proctor introduced agnotology as a discipline concerned with the production of ignorance; a discipline complementary to epistemology, that is, to the study of knowledge production.

[9]I can't treat here of the role of propaganda and media in the production of biased conceptions, and how this media device is used in generating fabricated doubt (cf. Ellul, 2014 [1965]).

One of the significant aims of the program for the production of ignorance is "to keep the controversy alive" (Leite, 2014, p. 181). It is based on a combined antiscientific strategy of (1) discrediting (producing false beliefs about) inconvenient scientific data, by producing other more favorable data (under different initial conditions, with different statistical methods etc.); and (2) defaming the scientific reputations of those researchers who present scientific evidence verifying the undesirable consequences of large scale application of particular techniques promoted by multinational corporations. The evident aim of "to keep the controversy alive" is to impede regulation, prohibitions and restrictions, or to postpone them as long as possible, while maintaining the state of technological exception. The fact that, in 125 cases brought against tobacco industry between 1954 and 1979 there was not a single conviction (cf. Leite, 2014, p. 182), proves the advantage for the corporations of maintaining the state of technological exception. It also shows that, despite all the impediments and restrictions that, from the 1990s on, were imposed by the courts on the tobacco industry and smokers, regulations were not able to be introduced, for example, concerning the composition of the substances in cigarettes. This is an evident sign that regulatory bodies have incorporated tobacco activities into the technological order - by limiting but not essentially modifying them, and that social inhibitions against using tobacco have to a certain extent been neutralized in the service of economic interests.

The same antiscientific practice reappears in more recent campaigns of the petroleum industry (from 1995 on) that aim to attack the Intergovernmental Panel on Climate Change (IPCC) (cf. Leite, 2014, p. 180-1), by producing doubts about the anthropic effects of climate change, in particular, the contribution of fossils fuels to climate change. However, recent discussions about the birth of a new geographical era - the Anthropocene - in the evolution of the Earth may be an index or an indication that technological environment has attained planetary scale, and that technology now may represent an important factor having impact on the climate dynamic of the Earth. Here, in view of the failure of all international attempts until now to set out limits to the emission of greenhouse gases produced by burning fossils fuels, one can clearly see that the state of exception is persisting.

The same kind of deliberate fabrication of ignorance occurs in connection with transgenic agriculture by the six companies - Monsanto (United States), Syngenta (Suisse), Dupont (United States), Basf (Germany), Bayer (Germany) and Dow (United States) (cf. Thuswohl, 2013, online) - that dominate the production of GMOs. They use a double strategy for the deliberate production of error. It involves, on the one hand, systematically conducting research aiming to challenge scientific data on the harm to environment and health caused by the extensive use of agrochemicals in agriculture and of on-going ingestion of GMOs; and, on the other hand, promoting campaigns to defame individual researchers, and even in some cases taking legal action, citing their intellectual property rights under patents, to prevent scientific research on the health and environmental effects of using transgenic plants and consuming their products.

The dissemination of fabricated ignorance deploys rhetorical skepticism against scientific organized skepticism. Instead of employing doubt to promote knowledge, it employs it to discredit scientific knowledge in order to disseminate ignorance. The skeptical procedure of doubting is then used with the deliberate aim of deceiving or defrauding. In contrast, the conjectural (hypothetical) and fallible character of scientific knowledge - that recognizes that scientific solutions are always subject to change (in view of knowing more, understanding better, and becoming less ignorant) in the light of rigorous empirical and conceptual critique and more severe tests based on scientific method - is linked with organized skepticism, one of the component values of the scientific ethos (cf. Merton, 2013, chap. 7) that is one of the outstanding institutional features of science. Applying the scientific method thus presupposes the acceptance of scientific ethos - a set of values of which the value of impartiality is part - and the practice of the virtues of honesty and sincerity that accompany the scientific use of method (practiced along with the values of organized skepticism), and respect for the moral imperatives of not lying about empirical data and evidences, of not using fraudulent research data, of not plagiarizing, etc. Thus, it presupposes certain precise moral commitments that are denied by the procedure of production of ignorance.

Significantly, creating doubt about scientific certifications, discrediting the legitimacy of security norms indicated by this certification, and postponing the introduction of regulations, maintains the state of technological exception and expands the range of health and environmental consumed facts. The aim is clear; the means employed to obtain it is immoral.

Ignorance and Ethics

The third antiscientific practice consists in de-legitimating scientific decisions, ignoring warnings about risks and harms to health and environmental occasioned by their technical applications, in order to keep the application in place where the state of technological exception can be maintained. Operating here, the consequence of transgression of the scientific ethos and of the immorality of the employed means, is a sort of disguising of knowledge or feigning of ignorance. However, to continue the use of agrochemicals, chemical fertilizers, and transgenic seeds locally in certain countries when this is already prohibited based on scientific certification of their extensive genetic, teratological, mutational effects etc. - for example, the herbicide 2,4-D (cf. Neumeister, 2014) - means that the corporations cannot claim ignorance of the consequences of their decisions of applying technical solutions with such known potential of consequences. Proposals to prohibit planting transgenic seeds and using agrochemicals that are based on scientific reports that recognize the danger of these effects, and also their contribution to the present ecological crises. Thus, to continue applying technical solutions, where there is compelling scientific certification that there are harmful health and environment consequences and residual effects, is to bring these harms about deliberately; this can be considered a criminal action. Furthermore, keeping in mind the co-evolution of human beings and their environment (cf. beginning of section 3 above), and the dependence of human survival on a sustainable environment, actions that bring about these harms, since they are deliberate, may be considered to be crimes against humanity: they modify the environment in such a way as to extend to future generations the harms caused by extensive application of agrochemicals, thereby decreasing the chances of survival of the human species.

This brings us back to the beginning of the article, to the problem of ethical responsibility for using such technological practices. Ethical responsibility demands an understanding of the "relation between personal freedom and the stable order" in view of the values adopted and of the objective knowledge at hand, both of which - values and knowledge - compose all of experience. Looked at in this way, actions of corporations and industries that are decided by a directorate, composed of individuals, cannot be treated ethically as the actions of a collective juridical personality. Ethical responsibility demands individual (personal) freedom of decision. So from the point of view of ethical responsibility, that is, of the relation between personal freedom and the social order, members of a directorate are, first and foremost, individuals - presidents, directors, scientists, technicians, engineers - to be held individually responsible for the consequences of decisions in which they are implicated.

Finally, the whole antiscientific program - undermining scientific impartiality, deliberately fabricating ignorance, and continuing to use technologies whose harmful consequences have been confirmed by sound scientific research - deviates from the scientific ethos, i.e., from the complex of values that has animated modern science. All of its components are expressions of the antiscientific attitude that animates the programs of technoscience and technological innovation of the neoliberal economy, and in line with the neoliberal conception of the minimum state and its program of deregulation. Deviating from the scientific ethos in this way has the effect of annulling the impact of ethics and thus undermining the most basic source of sensitivity to the requirements for humanity survival.

ACKNOWLEDGEMENTS

An earlier version of this article was presented at the 18th International Conference of the Society for Philosophy and Technology (SPT-2013), Lisbon, Portugal, July 6, 2013. I thank Hugh Lacey for his valuable help in discussing the main points of this article and for helping to put it in right English language. This article is part of the Thematic Project "Origins and significance of technoscience: on relations among science, technology and society", supported by Fapesp.

REFERENCES

1. Agamben, G. *Qu'est-ce qu'un dispositif?* Paris: Payot & Ravages, 2007
2. Bachelard, G. *Le rationalisme appliqué*. Paris: Vrin, 1949
3. Beck, U. *Sociedade de risco: rumo a uma outra modernidade*. São Paulo: Editora 34, 2010 [1986].
4. Bourg, D. Qu'est-ce qu'un risque environnemental? In: Kermisch, C. & Hottois, G. (Ed.). *Techniques etphilosophies des risques*. Paris: Vrin, 2007. p. 123-38
5. Cairns, H. & Cairns, H. (Ed.). *Plato: the collected dialogues*. Princeton: Princeton University Press, 1961
6. Cattani, A. D. & Holzmann, L. (Org.). *Dicionário de trabalho e tecnologia*. Porto Alegre: Zouk, 2011
7. Chazal, G. Dimension anthropologique du risque technique. In: Kermisch, C. & Hottois, G. (Ed.). *Techniques et philosophies des risques*. Paris: Vrin, 2007. p. 57-69

8. Dupas, G. *O mito do progresso ou progresso como ideologia*. São Paulo: Editora Unesp, 2006
9. Dusek, V. *Filosofia da tecnologia*. Tradução L. C. Borges. São Paulo: Loyola, 2009
10. Ellul, J. The technological order. *Technology and Culture*, 3, 4, p. 394-421, 1962
11. Feenberg, A. Ciencia, tecnología y democracia: distinciones y conexiones. *Scientiae Studia*, 7, 1, p. 63-81, 2009
12. Fressoz, J. B. *L'apocalypse joyeuse. Une histoire du risque technologique*. Paris: Seuil, 2012
13. Friedmann, G. *7 estudos sobre o homem e a técnica*. São Paulo: Difel, 1968
14. Holzmann, L. Automação. In: Cattani, A. D. & Holzmann, L. (Org.). *Dicionário de trabalho e tecnologia*. Porto Alegre: Zouk, 2011. p. 56-9
15. Keynes, J. M. *A treatise on probability*. New York: Harper, 1962 [1921] [Links].
16. Kermisch, C. & Hottois, G. (Ed.). *Techniques et philosophies des risques*. Paris: Vrin, 2007
17. Lacey, H. *A controvérsia sobre os transgênicos*. Aparecida/São Paulo: Ideias & Letras, 2006
18. Lacey, H. & Mariconda, P. R. The eagle and the starlings: Galileo's argument for the autonomy of science. How pertinent is it today? *Studies in History and Philosophy of Science*, 43, p. 122-31, 2012
19. Leite, J. C. Controvérsias científicas ou negação da ciência? A agnotologia e a ciência do clima. *Scientiae Studia*, 12, 1, p. 179-89, 2014
20. Martins, H. *Experimentum humanum*. Civilização tecnológica e condição humana. Belo Horizonte: Fino Traço, 2012
21. Mariconda, P. R. O controle da natureza e as origens da dicotomia entre fato e valor. *Scientiae Studia,* 4, 3, 453-72, 2006
22. Mariconda, P. R. & Ramos, M. de C. Transgênicos e ética: a ameaça à imparcialidade científica. *Scientiae Studia,*1, 2, p. 245-61, 2003
23. Merton, R. K. *Social theory and social structure*. Glencoe: The Free Press, 1957

24. Mumford, L. *A cidade na história: suas origens, transformações e perspectivas*. São Paulo: Martins Fontes, 2008 [1961].
25. Neumeister, L. *The risks of the herbicide 2,4-D*. Testbiotech/ GeneWatch UK/Pesticides Action Network (PAN) Europe. January 2014. Available at: <http://www.genewatch.org/uploads/f03c6d66a9b354535738483c1c3d49e4/Risks_of_herbicide_2_4_D.pdf>. Accessed: 20 Sept. 2014
26. Popper, K. R. *Logic of scientific discovery*. New York: Harper & Row, 1965
27. Proctor, R. N. *Cancer wars: how politics shapes what we know and don't know about cancer*. New York, 1995
28. Proctor, R. N. & Schiebinger, L. (Ed.). *Agnotology - the making and unmaking of ignorance*. Palo Alto, 2008
29. Putnam, H. *The collapse of the fact/value dichotomy and other essays*. Cambridge/London: Harvard University Press, 2004
30. Thuswohl, M. Grupo de seis empresas controla mercado global de transgênicos. *Repórter Brasil,* 2013. Available at: <http://reporterbrasil.org.br/2013/11/grupo-de-seis-empresas-controla-mercado-global-de-transgenicos-2/>. Accessed: 20 Aug. 2014

Chapter 6

Measuring Household Vulnerability to Climate-Induced Stresses in Pastoral Rangelands of Kenya: Implications for Resilience Programming

Francis EO Opiyo[1,2], Oliver V Wasonga[1,3], and Moses M Nyangito[1]

[1]Department of Land Resources Management and Agricultural Technology, University of Nairobi, Kabete, Nairobi 00625, Kenya

[2]Dryland Development Centre, United Nations Development Programme (UNDP), United Nations Office in Nairobi, Nairobi 00100, Gigiri, Kenya

[3]German Institute for Tropical and Subtropical Agriculture (DITSL), Steinstraße 19, Witzenhausen 37213, Germany

ABSTRACT

This study uses statistical and econometric tools to measure households' vulnerability in pastoral rangelands of Kenya. It considered 27 socio-economic and biophysical indicators obtained from 302 households' in-depth interviews to reflect climate vulnerability components: adaptive capacity, exposure and sensitivity. The theoretical framework used combines exposure and sensitivity to produce potential impact, which was then compared with adaptive capacity in order to generate an overall measure of vulnerability. Principal component analysis (PCA) was used to develop weights for different indicators and produce a household vulnerability index (HVI) so as to classify households according to their level of vulnerability. In order to understand the determinants of vulnerability to climate-induced stresses, an ordered probit model was employed with predictor variables. The results show that 27% of households were highly vulnerable, 44% were moderately vulnerable and 29% of households were less vulnerable to climate-induced stresses. Factor estimates of the probit model further revealed that the main determinants of pastoral vulnerability are sex of household head, age of household head, number of dependents, marital status, social linkages, access to extension services and early warning information, complementary source of income, herd size and diversity, herd structure, herd mobility, distance to markets, employment status, coping strategies and access to credit. Therefore, policies that address these determinants of vulnerability with emphasis on women's empowerment, education and income diversifications are likely to enhance resilience of pastoral households.

BACKGROUND

Vulnerability, commonly defined as the propensity or predisposition to be adversely affected, has been studied as a composite of adaptive capacity, sensitivity and exposure to hazards (Adger and Kelly [1999]; Kelly and Adger [2000]; McCarthy et al. [2001]; Intergovernmental Panel on Climate Change IPCC [2001]; Adger [2006]; Füssel [2007]; Paavola [2008]; Yuga et al. [2010]). Adaptive capacity is the ability of people to cope with or adjust to the changing context and is explained by socio-economic indicators. Sensitivity is the ability of a system to be affected, and exposure is the incidences of events (Kasperson et al. [1995]; Adger [2006]; Paavola [2008]). Vulnerability is thus comprised of risks or a chain of risky events that households confront in pursuit of their livelihoods, the sensitivity of livelihood to these risks, the response or options that households have for managing these risks and finally the outcomes that describe the loss in wellbeing (Turner et al. [2003]).

On the other hand, resilience is seen as the ability to self-organize, learn and adapt to risk hazards (Carpenter et al. [2001]; Turner et al. [2003]). The United Nations International Strategy for Disaster Reduction (United Nations International Strategy for Disaster Reduction-UNISDR [2009]) defines resilience as the transformative process of a household or community exposed to hazards to resist, absorb, accommodate and recover from hazards in a timely and efficient manner. Turner et al. ([2003]) and Gallopın ([2006]) define hazards as threats to a system and comprised of stressors. For simplicity, the term stress is used in this study to denote a continuous or slowly increasing pressure (in this case drought), commonly within the range of normal variability. However, over the last decades, most of the scientific literature and discourses on vulnerability have concentrated on contributing to theoretical insights or measurements at the regional or national scale, with selected indicators for each region, and identifying resilience-building strategies that have implication for national and regional planning (Brooks et al. [2005]; Füssel[2007]; Hinkel [2011]). Yet, micro-level vulnerability analysis is an essential prerequisite for local-level planning and prioritization of resilience planning and strategies especially among the natural resource-dependent communities at risk to projected climate variability and changes (Callaway[2004]; Fraser et al. [2011]).While there is no superior scale of climate vulnerability analysis, recent studies by Yuga et al. ([2010]) and Marshall et al. ([2014]) have confirmed that micro-level analyses have hitherto largely been overlooked in favour of ecosystem-scale studies of biophysical vulnerability.

As observed by Deressa et al. ([2008]), climate change vulnerability analysis ranges from local or household level (Adger [1999]) to the global level (Brooks et al. [2005]; Intergovernmental Panel on Climate Change IPCC [2014]). However, the choice of scale is dictated by the objectives, methodologies and data availability. This present study focused on household-level vulnerability analysis. After all, it is by understanding, planning for and adapting to a changing climate that individual households can take advantage of opportunities to reduce risks associated with climate-induced stresses (Madu [2012]). Similarly, Klein et al. ([2007]) observed that vulnerability analysis to climate change is needed at the level that would enable policy makers to tackle climate change challenges with the precision that is necessary, particularly in the arid and semi-arid regions of Africa. Previous studies by Brooks et al. ([2005]) and Intergovernmental Panel on Climate Change IPCC ([2012]) concluded that the majority of households are particularly vulnerable to the impacts of climate change in the arid and semi-arid regions. Although the causes of such vulnerability are multi-dimensional, they are primarily due to widespread poverty, food insecurity, recurrent droughts, land degradation, inequitable land distribution and overdependence on rain-fed agriculture (Notenbaert et al. [2013]; Lo´pez-Carr et al. [2014]). Some of the other common factors also postulated to determine adaptive capacity, and therefore influence vulnerability at the household level, include access to resources, markets and infrastructure; household structure, gender, education and age; farm size; income and income diversity; access to community-based organizations, information, credit facilities, savings and loans; and health status among others. Stringer et al. ([2009]) argued that though all households in a community are exposed to risks associated with climate change and could potentially be rendered vulnerable, the poorer households are the most at risk of adverse impacts of climate variability and change. That notwithstanding, the proportion of households vulnerable to extreme weather events is perceived to be increasing in arid environments of eastern Africa, especially in Kenya, and there is an uncertain degree to which the population or the system is becoming susceptible and unable to cope with hazards and stresses, including the effects of climate change.

In Kenya, there is a consensus that projected climate change will worsen food security, mainly through increased extremes and temporal or spatial shifts (Eriksen and O'Brien [2007]; Herrero et al. [2010]; Sherwood [2013]). In fact, there is considerable potential impact of these global drivers of change on the production systems and resource-poor households who depend on them. Studies by Kabubo-Mariara ([2009]) and Silvestri et al. ([2012]) revealed that extreme weather events such as prolonged dry spells and intense rainfall are already affecting rural communities in parts of arid and semi-arid lands (ASALs) of Kenya. At present, nearly 30% of the total human population resides in the

ASALs, which cover approximately 88% of the country's land mass, and hold almost 70% of the total national livestock herd. However, large proportions of pastoralists who reside in ASALs are believed to be at risk of food production deficit, with potential declines in pasture and water availability (Opiyo et al. [2011]), all exacerbated by extreme climate events. Hence, the purpose of this paper is to identify the determinants of vulnerability and measure micro-level vulnerability of pastoralists in arid rangelands of Kenya. The vital information that is obtained from household-level vulnerability analysis is presently lacking in national and regional level assessments. Deressa et al. ([2008]), Pearson et al. ([2008]) and Sherwood ([2013]) show that vulnerability contexts are diverse for different multiple spatial scales, and therefore this approach can contextualize how climate variability and change affect pastoralist livelihoods.

This study was thus carried out to investigate households' vulnerability to climate variability and change to climate-induced stresses in a case study of a pastoralist rangeland of Kenya. The study identifies some of the determining factors for vulnerability based on certain household social, economic and environmental (biophysical) characteristics. The household was selected as the main unit of analysis because major decisions about adaptation to climate-induced stresses and livelihood processes are taken at that level (Thomas [2008]). Nevertheless, households are connected to the wider community, which can greatly influence their decision-making processes in relation to use of particular productive resources. The findings of this study can therefore be useful for targeting interventions, priority setting and resource allocations at micro-level. Complemented with studies analysing climate change impacts and findings from country-level adaptive capacity, governmental policy can be informed. At the same time, the uncertainty associated with climate variability and change demands an approach that prepares rural households without relying on detailed climate projections. The study therefore focused on the adaptive capacity of households for wider resilience programming. Moreover, the findings would assist in resources allocation and determination of resilience investment opportunities that are likely to increase the adaptive capacity of the vulnerable households. The findings in particular suggest potential avenues for research that may further enhance understanding of household vulnerability to climate stresses in the arid rangelands of Kenya.

Study Area

The study was conducted in Turkana County, northwestern rangelands of Kenya (Figure 1). The county lies between longitude 34° 30′ and 36° 40′E and latitude 1° 30′ and 5° 30′ N, covering an area of 77,000 km² with an estimated human population of 855,399 according to the 2009 Kenya housing and population census report. The county is one of the poorest in Kenya with an estimated 87.5% of the population living below the absolute poverty line (Kenya National Bureau of Statistics KNBS [2013]). In Lokichoggio, Kakuma and Oropoi divisions where the household survey was undertaken, poverty is exacerbated by the harsh environment, poor infrastructure and low access to basic services, in addition to other underlying causes of poverty that are experienced in northern Kenya. About 96% of the study area falls into arid and very arid eco-climatic zones (Jaetzold and Schmidt [1983]). In the study area, because of aridity, managing short-term climatic fluctuations as well as adapting to long-term changes is critical in sustaining livelihoods. According to the most recent household economy approach assessments, pastoralism is the predominant livelihood accounting for over 55% of the population; 15.6% agro-pastoral, 8% fisher folks and 8.4% are in the urban/peri-urban areas (Household Economy Approach HEA [2012]).

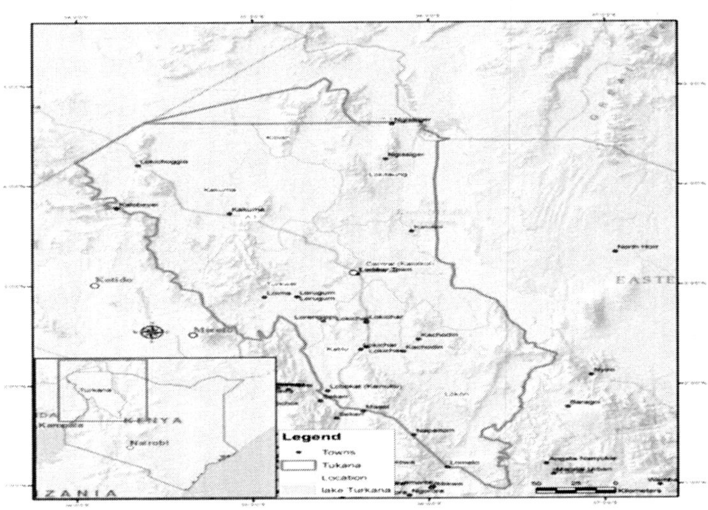

Figure 1: Location of the study area.

Rainfall in the study area exhibits a bimodal pattern, and distribution is unreliable and erratic in both space and time, with an average annual rainfall of 120 mm in the east to over 200 mm in the northwest. Figure 2 shows the annual rainfall variability of the study area. The main rainy season at local level between March 1950 and July 2012 the study area received very low and variable rain. In the past decades, major widespread drought events were experienced in 1980 to 1984, 1990 to 1995, 1999 to 2000, 2008 to 2009 and 2010 to 2011 within the study area. The last drought crisis in the Horn of Africa is estimated to have affected over 13 million people, including 3.75 million Kenyan, especially in the ASALs (UN-OCHA [2011]). The daily temperature ranges are extremely high with an average daily range varying between 24°C and 38°C with a mean of 30°C. The vegetation in this area is mainly scattered *Acacia* bush and a cover of annual herbaceous plants. The density of the woody plants such as *Acacia reficiens* and *A. mellifera* increases on hilly ground.

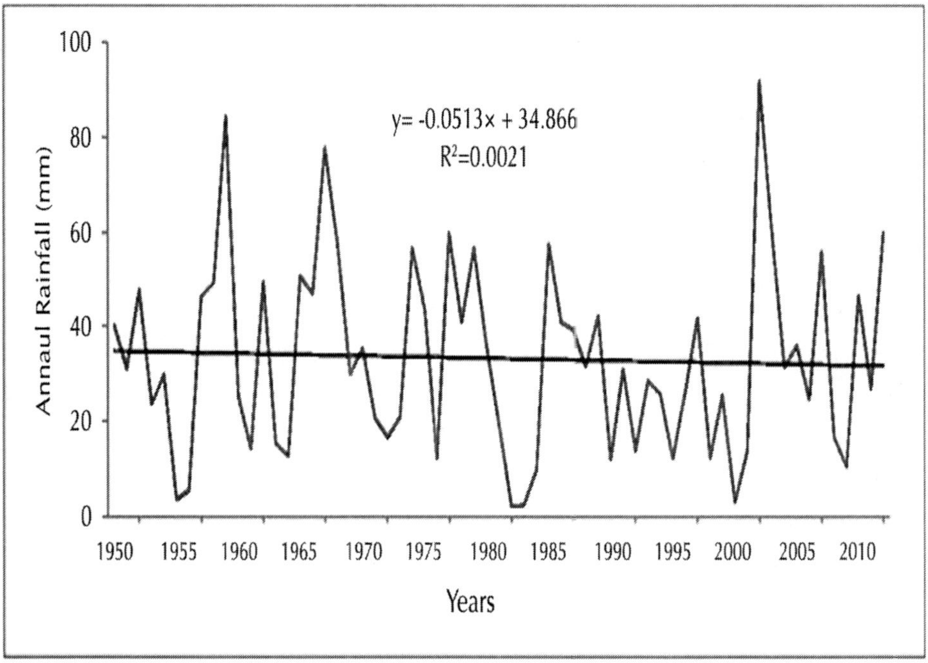

Figure 2: Annual rainfall variability for period 1950 to 2012 in the study area.

The study area is inhabited by the Turkana pastoralists whose major occupation is extensive livestock production (Watson and van Binsbergen [2008]).The livestock species kept include camels, cattle, sheep, goats and donkeys, which have different forage and water requirements and variable levels of resilience during drought episodes. In Turkana, droughts, famine, epidemics and flash floods are also very common occurrences. In most instances, these shocks and stresses are associated with climatic variability and change (Notenbaert et al. [2013]). The outcome of these climatic stresses has been high livestock mortality, loss of property, displacement of people and resources conflicts (Schilling et al. [2012]). In general, by weakening the productivity and functioning of livelihood resources, climatic stresses aggravate the vulnerability of households dependent on these resources. However, Turkana pastoralists are known to be highly mobile with no fixed residence (McCabe et al. [1988]), but their regular patterns of movement often get disrupted by conflicts and disease outbreaks (Hendrickson et al. [1998]; Schilling et al. [2012]).

METHODS

Data Collection

A total of 302 pastoralist households were selected for interviews a using semi-structured questionnaire to elicit responses between June and July 2012. Systematic purposive sampling was used to select drought-prone divisions in Turkana County which include Lokichoggio, Kakuma and Oropoi. However, the specific kraals (homesteads)/ locations where households were selected based on accessibility of the area. Consequently, 10 locations/kraals were selected from which the sample households were selected randomly proportional to population size. Table 1 shows the sampled households from each of the 10 kraals/ locations. A semi-structured questionnaire was used to interview the household heads. The questionnaire was divided into the following: demographic and economic household characteristics, livestock and crops production, access to extension services, credit access, hazards occurrence, perception level, adaptation strategies pursued, different coping strategies, level of resilience and other relevant information.

Table 1: Sampled households by Kraal

Location/Kraals	Number of households	Sampled household	Percentage of respondent
Mogila	1,536	20	6.62
Lokichoggio	1,868	9	2.98
Lopusiki	649	1	0.33
Lopwarin	277	11	3.64
Songot	459	19	6.29
Nanam	748	38	12.58
Kakuma	12,767	33	10.93
Letea	4,957	56	18.54
Kalobeyei	2,577	39	12.91
Pelekech	1,863	76	25.17

Kraal used herein to refer to a traditional pastoralists village of huts, typically enclosed by a fence. (Source: Authors' compilation). Opiyo *et al.*

Opiyo *et al. Pastoralism* 2014 **4**:10, doi:10.1186/s13570-014-0010-9

In addition, *in situ* rainfall and temperature datasets relevant for this study were obtained from Kenya Meteorological Services, National Drought Management Authority (National Drought Management Authority - NDMA [2013]) and County government officers in Lodwar, Turkana to strengthen some of the study findings. The STATA software was used to analyse the data obtained.

Analysis of Households' Vulnerability

There are various ways of analysing vulnerability, namely, socio-economic, biophysical and an integrated approach, which unites both socio-economic and biophysical factors. The socio-economic vulnerability assessment approach focuses on the socio-economic and political status of individuals or groups. Individuals in a community vary in terms of education, gender, wealth, health status, access to credit, access to information and technology, formal and informal (social) capital and political power, which are responsible for variations in vulnerability levels (Füssel[2007]; Deressa et al. [2008]). Consequently, vulnerability is considered to be a starting point or a state that exists within a system before it encounters a hazard event (Kelly and Adger [2000]). In this regard, vulnerability is shaped by society as a result of institutional and economic changes. The socio-economic approach focuses on identifying the adaptive capacity of individuals or communities based on their internal characteristics. One major limitation of this approach is that it focuses only on variations within society, but in reality, societies vary not only due to socio-political factors but also because of environmental or biophysical factors. The socio-economic approach does not account for the natural resource bases which have the potential to counteract the negative impacts of environmental shocks. For example, areas with easily accessible underground water can better cope with drought by utilizing this resource, compared to areas without it (Deressa et al. [2008]).

The second commonly used approach is biophysical that attempts to assess the level of damage that a given environmental stress causes on both social and biological systems. It is sometimes known as an impact assessment. The emphasis is on the vulnerability or degradation of biophysical conditions (Liverman [1990]). This is a dominant approach employed in studies of vulnerability to natural hazards and climate variability and change (Hewitt [1995]). Füssel ([2007]) identified this approach as a risk-hazard approach. The biophysical approach, although very informative, also has a major limitation that assessment of biophysical factors is not a sufficient condition for understanding the complex dynamics of vulnerability. This approach also neglects both structural factors and human agency in producing vulnerability and in coping or adapting to it. The approach overemphasizes extreme events while neglecting root causes and everyday social processes that influence differential vulnerability (Liverman [1990]; Hewitt [1995]; Pulwarty and Riebsame [1997]). The third approach is the integrated vulnerability analysis, which combines both the socio-economic and biophysical factors. This approach includes all the internal state of vulnerability and the external situation. This analytical approach was applied by Madu ([2012]) in agro-ecological based household vulnerability analysis in Ethiopia and by Deressa et al. ([2008]) in regionally based vulnerability analysis.

This present study replicates an integrated vulnerability approach to develop vulnerability indices for each household as proposed by Madu ([2012]) and adopted by Tesso et al. ([2012]) in Ethiopia. In this research, it is assumed that households with higher adaptive capacity are less sensitive to impacts of climate-induced stresses, thus keeping the level of exposure constant. The integrated assessment approach combines both socio-economic and biophysical approaches to determine vulnerability. The vulnerability index development is given as developed by the Intergovernmental Panel on Climate Change IPCC ([2012]) that vulnerability is seen as the net effect of adaptive capacity (socio-economic) and sensitivity/exposure (biophysical):

$$\text{Vulnerability}=(\text{Adaptivecapacity})-(\text{Sensitivity}+\text{Exposure}) \quad (1)$$

When the adaptive capacity of the household exceeds that of its sensitivity and exposure, the household becomes less vulnerable to climate change impacts and the reverse is also true. As explained in the foregoing, each set (adaptive capacity, sensitivity and exposure) is composed of different variables. The model specification further looks like

$$V_i = (A_1X_{1J} + A_2X_{2J} + \ldots + A_{2n}X_{nJ}) - (A_{n-1}Y_{IJ} + A_{n+2}Y_{2J} + \ldots + A_{n+n}X) \quad (2)$$

where V_i is the vulnerability index, while X_s are elements of adaptive capacity, and Y_s are elements of exposure and sensitivity. The values of X and Y are obtained by normalization using their mean and standard errors. For instance, $X_{IJ}=(X_{IJ}-X_{1*})/S^{1*}$, where X_{1j*} is the mean of X_{1j} across the different households, S_1* is its standard deviation. X_1 is the principal component result of factors. In this regard, the first principal component of a set of variables is the linear index of all the variables that captures the largest amount of information common to all the variables. The whole matrix of variables of adaptive capacity (X_{ij}) and variables of exposure and sensitivity (Y_{ij}) appears as follows:

$$X_{ij} / Y_{ij} = \begin{pmatrix} (X_{11} + X_{12} + \ldots + X_{2n}) - (Y_{11} + Y_{12} + \ldots + Y_{2n}) \\ (X_{m1} + X_{m2} + \ldots + X_{mn}) - (Y_{m1} + Y_{m2} + \ldots Y_{mn}) \end{pmatrix} \quad (3)$$

The *i* and *j* in the foregoing notation imply the number of rows (in this case is the 302 individual households) and the number of columns (27 variables of adaptive capacity, exposure and sensitivity). In Equation 4, the A_s is the first component score of each variable computed using principal component analysis (PCA) in STATA. Finally, the vulnerability index of each household is obtained using Equation 4 as follows:

$$V_i = \begin{pmatrix} A_1 \\ A_{12} \\ \cdot \\ A_n +_n \end{pmatrix} \times \begin{pmatrix} (X_{11} + \ldots + X_{2n}) - (Y_{11} + \ldots + Y_{2n}) \\ (X_{m1} + \ldots + X_{mn}) - (Y_{m1} + \ldots + Y_{mn}) \end{pmatrix} \quad (4)$$

In calculating the direction of relationship in vulnerability indicators (that is, their sign), a negative value was assigned to both exposure and sensitivity. The justification is that households which are highly exposed to climate shocks are more sensitive to damage, assuming constant adaptive capacity. The implication is that a higher net value indicates lesser vulnerability and vice versa. However, in creating the indices, the scale of analysis is important. As noted by Tesso et al. ([2012]) from Deressa et al. ([2008]), vulnerability analysis ranges from local or household level to the global level (Brooks et al. [2005]). The choice of scale is dictated by the objectives, methodologies and data availability. In this study, the households were classified into three categories based on the value of their vulnerability index, which puts households into highly vulnerable, vulnerable and less vulnerable categories. However, the index computed is not based on the thresholds or presents an absolute value. It is a relative measure, representing the households' own perception of how they have been coping in the past compared to other households.

Factors Influencing Household Vulnerability

Notenbaert et al. ([2013]) note that many factors contribute to vulnerability, and these factors undermine capacity for self-protection, block or diminish access to social protection, delay recovery or expose some households to greater or more frequent hazards than other households. The analysis was performed using ordinal logistic regression analysis. The ordinal logit model is used when the outcome variable is categorized in an ordinal scale, as in this case where vulnerability is ordered as (1) highly vulnerable, which implies households for whom the difference between adaptive capacity and sensitivity/exposure is significantly negative; (2) moderately vulnerable, which means that households for whom the difference between adaptive capacity and sensitivity/exposure is nearly zero; and (3) less vulnerable, which means that the difference between adaptive capacity and exposure/sensitivity is significantly positive. In this study, sensitivity of households to climate-induced stresses is represented by its associated impacts, i.e., shortage of food, loss of water and pasture resources and conflicts faced by those households. In the case of exposure, since all households are assumed to be located in the same environment, exposure is almost uniform across the respondents residing in the study area.

This model is particularly useful in that it can show movement between vulnerability groups, explaining who moves in and out of vulnerability. Following Greene ([1997]), the reduced form of the ordinal logit model is given as

$$Y_j^* = X_j^1\beta + U_{lj} \tag{5}$$

where *Y* is the level of vulnerability and involves ordered outcome, that is, *Y*=1 was given to households that have a high level of vulnerability as observed by the negative value of adaptive capacity minus sensitivity/exposure; *Y*=2 was given to households having adaptive capacity nearly equal to their sensitivity/exposure; and *Y*=3 was given to households having their adaptive capacity exceeding their sensitivity and exposure. Y^* is the given state of vulnerability. The X_{ij} are the explanatory variables determining vulnerability level. The independent variables included in the model were sex and age of the household head, experience in the study area, household size and education level of the household head, dependency, marital status, social linkages and visits by extension officers, access to early warning information, non-farm income, herd size, herd structure, access to markets, property regimes, access to remittances, employment, coping strategies, herd diversity, credit access, herd mobility, climate change, experience in increased temperature, drought, floods, and wind, and natural hazards encountered in 5 years. β_s are parameters estimated and U_{ij} is the disturbance term. Y^* is unobserved, but what was observed in this study is

$$Y=1 \text{ if } Y^* \leq \mu 2 \tag{6}$$

$$Y = 2 \text{ if } \mu 2 < Y^* \leq \mu 3 \tag{7}$$

$$Y = 3 \text{ if } \mu 3 < Y^* \tag{8}$$

Given the cumulative normal function Φ $(\beta' x)$, the probabilities can be shown, thus,

Prob [$y=1$ or highly vulnerable] $= \Phi\ (-\beta' x)$,

Prob [$y=2$ or neutral level of vulnerability] $= \Phi\ (\mu 2 - \beta' x) - \Phi\ (\mu 3 - \beta' x)$

Prob [$y=3$ or less vulnerable] $= 1 - \Phi\ (\mu 3 - \beta' x)$

RESULTS

Hazards Reported by Households

The results show that 49% of households experienced drought over the past three decades, with 27.1% of the respondents reporting livestock disease outbreak, 19% mentioning cross-border inter-community conflicts and 4% citing riverine flash floods as the main hazard (Table 2). However, it was difficult for the respondents to differentiate between threats and hazards, which required clarification of the differences. Drought events were reported to be frequent hazards in the area and had devastating impacts on household livelihoods, pasture and water, which escalates the area's chronic conflicts, insecurity and food insecurity.

Table 2: Hazards identified by respondents

Hazard	Description of hazards	Number of respondents	Percentage of respondents

Floods	Flash floods along ephemeral rivers	13	4.3
Droughts	Widespread frequent droughts	148	49.0
Disease outbreak	Livestock diseases outbreak are common	82	27.1
Others i.e. conflicts, fire	Inter-community conflicts along the borders over resources	59	19.5

United Nations ([2004]) defines a 'hazard' broadly as "a potentially damaging physical event, phenomenon or human activity that may cause the loss of life or injury, property damage, social and economic disruption or environmental degradation". (Source: Authors' compilation).

Opiyo *et al.*

Opiyo *et al. Pastoralism* 2014 **4**:10, doi:10.1186/s13570-014-0010-9.

Inter-community and ethnic conflict related to scarce grazing land and water resources was reported to have increased inter-tribal animosity, often resulting in armed violent conflicts, which are predatory in nature and much more destructive. Households stated that violent conflicts regularly lead to heavy losses of lives and livelihoods, undermining human and livestock population mobility, as well as development efforts. Study findings show that outbreaks of *peste des petit ruminants* (PPR) locally referred to as *lomoo* is considered as the major hazard, having the highest impact on small ruminants compared to other endemic livestock diseases in the area. On further probing with community animal health workers and veterinary officers, based on the symptoms described by the pastoralist respondents, other diseases were identified as *etome*(mange), *emadang'* (worms infestation), *lukoi* (contagious caprine pleuropneumonia), *lomeri* (lump skin disease), *lokichum* (heart water), *etune* (sheep/goat pox), *lotorebwo or lokipi*(trypanosomiasis), *logooroi* (hemorrhagic septicaemia) and *loukoi* (contagious bovine pleuropneumonia).

Socio-Economic and Biophysical Vulnerability

The social and economic variables contributing to vulnerability are summarized in Table 3. Findings show that more than 80% of the respondents had no basic primary education, while the majority are not able to read and write. This in turn reduces a household's ability to understand climatic information, access market and early warning information. Data on household size showed that 64% of those interviewed had more than five persons, with 43% respondents reported more than five dependents. Whereas the sex of the household head is likely to be critical for climate adaptation, the study revealed that more than 50% of households are headed by females. In addition, 54% of the respondents indicated that they do not have access to livestock extension services. The results imply that the vulnerability level of households to the frequently occurring climate-induced stresses is largely determined by gender and education level of the household head in the study area.

Table 3: Social, economic and environmental indicators and their effect on vulnerabilitylevel

Hypothesized variables	Percentage	Influence on vulnerabilitya
Social vulnerability variables		
Sex of HH head: female-headed	50.7	+
Age of HH head: 50+ years	26.6	–
Experiences in the area: less than 5 years	7.9	+
HH size: more than 5 persons	64.6	–
Educational level: no primary education	80.1	+
Dependents: more than 5 persons	43.8	–
Marital status: single (including divorced and widowed)	29.5	+
Linkages: having no social linkages	25.8	–
Visit by extension officers: no access to extension services	54.3	+
Access to early warning information: no access to the information	74.8	+
Economic vulnerability variables		
Non-farm income: have no non-farm income	35.5	–
Herd size: own less than 2 TLUs	18.3	+
Herd structure: no milking herd	72.8	+
Distance to market: more than 10 km away	39.4	–
Property regime: own private land	8.6	–

Access to remittances: no cash transfers	53	+
HH employed: no member of HH employed	83.1	+
HH coping strategies: more than 2 coping strategies	92.1	+
Livestock diversity: less than 2 livestock species	45.4	–
Credit access: having no access to credit at all	77.8	+
Mobility: able to move livestock freely	67.9	–
Environmental vulnerability variables		
Climate change: experiencing change	96.7	+
Temperature: experiencing increase	47.9	+
Drought: noticed increasing events	3.4	+
Flood: noticed change	4.1	+
Wind: noticed unusual change	11.6	+
HH facing more than 2 hazards in 5 years	72.5	–

[a]Positive sign indicates that the variable increases vulnerability, while negative sign means it reduces vulnerability. TLU, tropical livestock unit (1 TLU = 250 kg). (Source: Estimated from June to July 2012 household survey interviews in Turkana; Authors' compilation).

Opiyo *et al.*

Opiyo *et al. Pastoralism* 2014 **4**:10, doi:10.1186/s13570-014-0010-9.

More than 35.5% of the households had no diversified income sources and heavily depend on livestock as their source of income. Results show that only a few of the households were practicing dry land crop farming alongside livestock keeping. For the livestock owned, 18% of respondents had less than two tropical livestock units (TLUs[a]), and 72% of households had no milking herd during the interview period. Further, the result revealed that about 68% of households practice seasonal herd mobility and flexible resource use in the study area. The results show that 88% of households had none of its member formally employed, and 92% of respondents use more than two coping strategies to survive in this harsh climatic condition, despite the low access to credit, access to remittances and long distances to markets. Other economic concern mentioned was the high formal unemployment opportunities, especially for the youths in Turkana.

Footnote

[a]A TLU is the 250-kg live weight of any domestic herbivore (sheep or goat=0.1 TLU; donkey=0.4 TLU; cattle=0.7 TLU; camel=1.0 TLU).

Table 3 displays environmental, economic and biophysical factors hypothesized for climate-induced vulnerability. The results show that the sensitivity and exposure to climate variability in terms of those experiencing climate variability (96.7%) and people facing more than two hazards in 5 years (72.5%) have contributed negatively to the vulnerability level of households. Approximately 48% of the respondents noticed an increase in temperature. However, most of the environmental variables increase household's level of vulnerability to climate-induced stresses as indicated by the positive sign.

Measuring Household-Level Vulnerability

Table 4 presents the result of principal component analysis and its association with the social, economic and environmental variables. The factor scores (weights) of the first principle component analysis was positively associated with the majority of the indicators identified under adaptive capacity, exposure and sensitivity. Holding exposure and sensitivity constant, a negative index shows the household to have relatively lower adaptive capacity when compared to a household with a positive index value and vice versa.

Table 4: Factor score for the first principal component analysis

Factors	Factor Score
Social vulnerability variables	
Sex of HH head: female-headed	0.22917
Age of HH head: 50+ years	−0.25804
Experiences in the area: less than 5 years	−0.02906
HH size: more than 5 persons	−0.29837
Educational level: no primary education	0.039479
Dependents: more than 5 persons	−0.31599
Marital status: single (including divorced and widowed)	0.07224

Linkages: having no social linkages	−0.04414
Visit by extension officers: no access to extens on services	0.189862
Access to early warning information: having no access	0.19573
Economic variables	
Non-farm income: HH with no farm income	−0.34130
Herd size in TLU: own less than 2 TLUs	0.293813
Herd structure: no milking herd	0.100562
Distance to markets: more than 10 km away	−0.15870
Property regime: own private land	−0.06728
Access to remittances: no cash transfers	0.047677
HH employed: no member of HH employed	0.095173
HH coping strategies: more than 2 coping strategies	0.101335
Livestock diversity: less than 2 livestock species	−0.42800
Credit access: having no access to credit at all	0.178153
Mobility: able to move livestock freely	−0.14718
Environmental variables	
Climate change: experiencing change	−0.02276
Temperature: experiencing increase	0.081161
Drought: noticed increasing events	0.164723
Flood: noticed change	0.039066
Wind: noticed unusual change	0.259975
HH facing more than 2 hazards in 5 years	−0.06667

Abbreviations: Households (HH), Tropical livestock unit (TLU). Opiyo *et al.*
Opiyo *et al.* *Pastoralism* 2014 **4**:10, doi: 10.1186/s13570-014-0010-9

To compute the vulnerability index in Equation 2, indicators of adaptive capacity, which are positively associated with the first principal component analysis, and indicators of sensitivity and exposure, which are negatively associated with the principal component analysis, were used in this study. The variables considered in Equation 2 include sex of the household head, education level, marital status, access to extension services and early warning information, livestock ownership, herd structure, access to cash remittances, household employment status, coping strategies and access to credit. However, for the exposure and sensitivity, all the variables were considered in the analysis. This is because adaptive capacity is considered as positively contributing to the reduction of vulnerability, while exposure and sensitivity are negatively contributing to vulnerability reduction. The larger the factor score the more important is the variable and contributes more to the household's vulnerability.

The households were classified into three categories using the vulnerability index: less vulnerable are households that are in a vulnerable situation but can still cope; moderately vulnerable households are those that need urgent but temporary assistance in case of shock and stresses; and the highly vulnerable are those households that are almost at a point of no return. The result shows that the majority of households fall within the moderately vulnerable category, with 44% households having an index from −1.00 to 1.00. The less vulnerable households had an index of 1.1 to 3.0 and constitute 29%, while the highly vulnerable households had an index of −0.9 to −3.0 but are 27% of the total households sampled (Table 5 and Figure 3). Although there seems to be normal vulnerability index distribution in Figure 3, a keen look at the values showed a slight shift to the left. Statistically, the households' distribution is skewed toward the highly vulnerable in the illustration. In general, the results reveal high and moderate vulnerability levels of the pastoralist households in Turkana.

Table 5: Classification of community according to the range of their vulnerability index

Vulnerability category	Household situation	Vulnerability index	Percentage of households
Highly vulnerable	Emergency level HHs	−0.9 to −3.5	27
Moderately vulnerable	Needs urgent but temporary external assistance to recover	−1.0 to +1.0	44
Less vulnerable	In a vulnerable situation but still able to cope	+1.1 to +3.0	29
Total			100

(Source: Computed from household questionnaire interviews in June to July 2012).

Opiyo *et al.*

Opiyo *et al. Pastoralism* 2014 4:10, doi:10.1186/s13570-014-0010-9

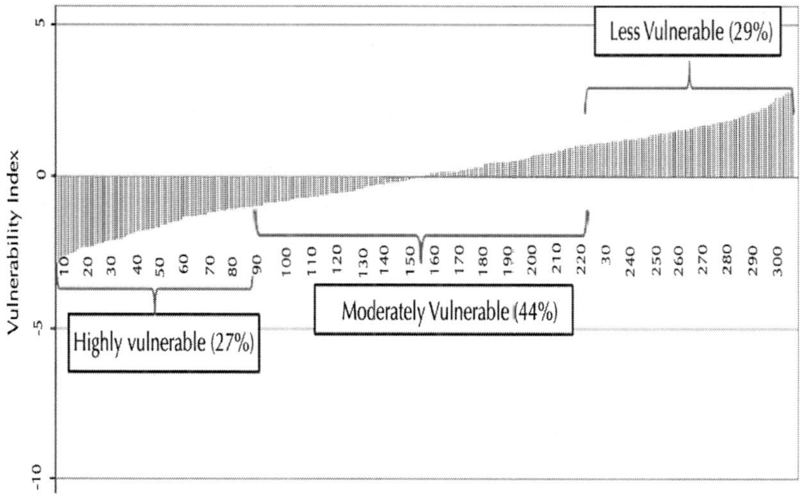

Figure 3: Vulnerability index by households in Kakuma, Oropoi and Lokichoggio divisions in Turkana.

Determinants of Vulnerability

The majority of highly vulnerable households were headed by females or someone with no primary level of education (Table 6). The highly vulnerable households also had more than five dependents, no access to early warning information, no milking herd and own less than two livestock species, and perceived changes in climate. By contrast, a household is likely to be less vulnerable when they are headed by a male, the household head is literate, not divorced or widowed, when they have access to extension services and early warning information, and own large and mixed-species herds. Similarly, households with access to cash remittances, with a member of the household in formal employment, and those with diverse coping strategies and access to credit facilities are reported to be less vulnerable to climate extremes. The results suggest that because of scarce resources during extreme climate events, households with high number of dependents and without any other livelihood diversification activities are likely to be more vulnerable than households with complementary sources of income and less dependents.

Table 6: Statistical description of model variables by vulnerability category

Vulnerability variables	Vulnerability category		
	Less vulnerable	Moderate vulnerable	Highly vulnerable
Sex of HH head: female	67 (44%)	62 (20%)	24 (45%)
Age of HH head: 50+ years	6 (41%)	33 (34%)	35 (25%)
Experiences in the area: less than 5 years	10 (16%)	10 (44%)	4 (4%)
HH size: more than 5 persons	27 (8%)	92 (22%)	76 (42%)
Educational level: no primary education	72 (45%)	107 (5%)	63 (55%)
Dependents: more than 5 persons	10 (47%)	58 (41%)	59 (33%)
Marital status: single (including divorced and widowed)	31 (42%)	41 (55%)	17 (46%)
Linkages: HH with no social linkages	18 (42%)	34 (52%)	26 (21%)
Visit by extension officers: no access to extension services	67 (17%)	65 (39%)	32 (20%)
Access to early warning information: having no access	77 (14%)	100 (9%)	49 (45%)
Non-farm income: HH with no farm income	5 (47%)	43 (33%)	58 (34%)
Herd size in TLU: own less than 2 TLUs	23 (39%)	17 (44%)	4 (31%)
Herd structure: no milking herd	72 (30%)	97 (23%)	51 (42%)
Distance to market: more than 10 km away	20 (44%)	51 (18%)	40 (27%)
Property regime: own private land	4 (26%)	13 (46%)	9 (40%)
Access to remittances: no cash transfers	58 (8%)	71 (36%)	46 (30%)
HH employed: no member of HH employed	81 (46%)	118 (15%)	67 (30%)
HH coping strategies: more than 2 coping strategies	82 (46%)	121 (50%)	66 (25%)
Livestock diversity: less than 2 species	5 (35%)	57 (35%)	75 (58%)

Credit access: having no access to credit at all	78 (46%)	108 (33%)	49 (17%)
Mobility: able to move livestock freely	42 (19%)	93 (41%)	70 (32%)
Climate change: experiencing change	43 (23%)	59 (26%)	38 (47%)
Drought: noticed increasing events	4 (44%)	3 (30%)	3 (21%)
Flood: noticed change	3 (33%)	7 (44%)	2 (29%)
Wind: noticed unusual change	11 (41%)	16 (25%)	7 (45%)
HH facing more than 2 hazards in 5 years	63 (40%)	99 (30%)	57 (26%)

HH, households; TLU, tropical livestock unit. Opiyo *et al.*

Opiyo *et al. Pastoralism* 2014 **4**:10, doi:10.1186/s13570-014-0010-9

As shown in Table 6, the majority of households in the moderately vulnerable category are the ones with less than 5 years' experience, either divorced or widowed household heads, household heads with no social linkages, household heads with no access to extension services, households who own less than two TLUs, own private lands, households which do not received any cash remittances, household heads with more than two coping strategies, households who practice mobility and are able to move freely with their livestock herd, and perceive climatic changes in the area.

The results of the ordered logistic regression model for all the single predictors' variables influencing a household's vulnerability are summarized in Table 7. A number of variables were statistically significant at 1% and 5% levels of significance and were important in influencing households' vulnerability to climate-induced stresses. The variables that showed significant influence on vulnerability include gender of the household head, age of the household head, size of the household, number of dependents, marital status, access to extension services and early warning information, complementary sources of income, herd diversity and structure, coping strategies and access to credit.

Table 7: Factors influencing household's vulnerability

Variables	Estimate	SE	OR	z	pvalue
Sex of HH head: female	−1.3828	0.2303	0.2509	−6.0049	<0.0001*
Age of HH head: 50+ years	1.5714	0.2704	4.8134	5.8106	<0.0001*
Experiences in the area: less than 5 years	−0.6187	0.399	0.5386	−1.5506	0.1220
HH size: more than 5 persons	2.2155	0.2651	9.1663	8.3585	<0.0001*
Educational level: no primary education	−0.1754	0.2684	0.8391	−0.6534	0.5140
Dependents: more than 5 persons	2.1434	0.2639	8.5281	8.1212	<0.0001*
Marital status: single (including divorced and widowed)	−0.4689	0.2349	0.6257	−1.9965	0.0468*
Linkages: having no social linkages	0.4232	0.2456	1.5268	1.7233	0.0859**
Visit by extension officers: no access to extension services	−1.0717	0.2237	0.3424	−4.7909	<0.0001*
Access to early warning information: having no access	−1.017	0.2535	0.3617	−4.0119	<0.0001*

Non-farm income: HH with no farm income	2.3332	0.2715	10.3108	8.5946	<0.0001*
Herd size in TLU: own less than 2 TLUs	−1.8785	0.3414	0.1528	−5.502	<0.0001*
Herd structure: no milking herd	−0.668	0.2434	0.5128	−2.7442	0.0064*
Distance to market: more than 10 km away	0.7694	0.2301	2.1586	3.3442	<0.0001*
Property regime: own private land	0.5666	0.375	1.7623	1.5111	0.1318
Access to remittances: no cash transfers	−0.2696	0.2161	0.7637	−1.2477	0.2131
HH employed: no member of HH employed	−0.6226	0.3337	0.5365	−1.866	0.0630**
HH coping strategies: less than 2 coping strategies	−1.0896	0.4268	0.3363	−2.5529	0.0112*
Livestock diversity: less than 2 livestock species	3.3875	0.3506	29.592	9.663	<0.0001*
Credit access: having no access to credit at all	−1.1825	0.2703	0.3065	−4.3753	<0.0001*
Mobility: able to move livestock freely	1.2984	0.2424	3.6635	5.3571	<0.0001*
Temperature: experiencing increase	−0.0412	0.2175	0.9597	−0.1892	0.8500

Drought: noticed increasing events	−0.2123	0.6269	0.8087	−0.3387	0.7350
Flood: noticed change	−0.1252	0.5225	0.8823	−0.2396	0.8108
Wind: noticed unusual change	−0.2462	0.3355	0.7818	−0.7338	0.4636
HH facing more than 2 hazards in 5 years	−0.0369	0.2409	0.9638	−0.1532	0.8784

SE, standard error; OR, odds ratio; *z*, score of two sample tests (basically a *t* test) though using the standard normal to calculate the *p* value and expressing the statistically significance levels at *5% and **10%. (Source: Ordinal logistic regression model: single predictors).

Opiyo *et al.*

Opiyo *et al. Pastoralism* 2014 **4**:10, doi:10.1186/s13570-014-0010-9

The higher odds ratios computed for age of the household head, household size, dependents, complementary income sources, herd diversity and mobility imply that these factors have more effects on the households' vulnerability level. But it should be noted that majority of the sampled population derived their means of subsistence from nomadic pastoral production system which are weather sensitive and therefore absence of non-agricultural income sources (salary or wages labour) which are not weather sensitive would be a significant determinant on vulnerability level as reflected in the odds ratio figures. Herd diversity and mobility allow households to spread risk and pursue various coping strategies. Herd mobility was observed to be a significant (at 5%) determinant of households' vulnerability to climate variability and change.

From the ordered regression results, 75% of female-headed households are more likely to be vulnerable to climate-induced stresses and shocks compared to the male-headed households. This might be because female-headed households normally face gender discrimination with respect to resources, rights, education, income and economic opportunities. Household age was also an important demographic factor determining how vulnerable a household could be. For example, households headed by person above 50 years of age are more likely to be vulnerable compared with the younger persons. Consequently, elderly household heads are probably worse off in terms of preparing strategies to cushion their families against adverse climatic stresses and impacts and likely to make them more vulnerable. Similarly, the more dependents a household has the more likely for it to be vulnerable since a larger proportion of household resources are directed to dependents who cannot contribute much toward household welfare. Further results also show that widowed or divorced single-headed families are 37.4% more likely to be vulnerable than families where both spouses are present. According to the study results, it was observed that livestock diversity had higher odds ratio of 29.592, which suggest that households with more than two livestock species are 28 times more likely to shift from highly vulnerability category to moderately vulnerability level.

Although 27 variables were hypothesized to be correlated with vulnerability, ordinal regression model result confirmed that only 17 factors were significant (at $p<0.1$ and $p<0.05$) in influencing households' vulnerability. Although not computed in this present study, it should be noted that different determinants have differential effects on a household's vulnerability levels.

DISCUSSION

Drought was considered by the respondents as one of the most frequent hazards in Turkana, in addition to other hazards such as conflicts, disease outbreaks and flooding. The pastoralist respondents maintained that frequent, prolonged drought events have increased in severity over the past decade or so. This perception corroborate Nicholson's ([2014]) assertion about drought events in northern Kenya, with rainfall being at least 50% to 75% below normal in almost half of the drought-stricken region and is likely to be linked to changing climatic conditions within the greater Horn of Africa. However, there is a great deal of uncertainty and tension about when, where and how much climate change will happen. Even less is known about the actual impacts of these perceived changes (Kabubo-Mariara [2009]; Notenbaert et al. [2013]; Intergovernmental Panel on Climate Change IPCC [2014]) to the dry land communities. It is argued that due to the frequency of shocks in the study area, adaptive or coping capacity and mechanisms adopted by vulnerable households could well have equal or larger ranges to that of exposure and sensitivity. As discussed earlier on vulnerability, the interaction of environmental (biophysical) forces determines exposures and sensitivities, and various social, cultural, political and economic forces shape adaptive capacity (Yohe and Tol [2002]; Turner et al. [2003]; Skjeflo [2013]).

In the study area just like the rest of northern Kenya, pastoralist households largely bear the brunt of negative impacts of extreme climate events like drought, which include increased poverty, water scarcity, resource-based conflicts, disease outbreaks and food insecurity. The majority of people in Turkana live below the absolute poverty line (an estimated 87.5% of the population) with more than 50% heavily relying on food aid and safety net programmes from year to year. The people who are already poor in this remote part of the country are struggling to cope with the added burden of increasingly unpredictable weather, which is triggered by climate variability and change. Evidence shows that Turkana is constrained by the harsh climatic conditions and remoteness, coupled with poor infrastructure and low access to essential services (Republic of Kenya [2012]) Previous studies (Thornton et al. [2006]; Kabubo-Mariara [2009]) in the region confirmed that it is becoming increasingly difficult for households to bounce back from ever-changing, inconsistent weather affecting their livelihoods, and many have been forced to pursue other livelihoods and coping mechanisms that only increase the cycle of vulnerability.

This study has shown that female-headed households, households with experience of less than 5 years in the area, household heads with no primary level of education and households headed by divorced and widowed persons, with no access to extension services and early warning information, in particular, are disproportionately likely to be affected by climate stresses and variability. In times of climate stresses and shocks like drought, these categories of households tend to have fewer options to find other ways of making a living, because their very low levels of literacy reduce their opportunities in coping mechanisms such as wage employment. Similarly, female or divorced and widowed household heads are likely not to be empowered enough in pastoral communities to make household decisions (Nabikolo et al. [2012]) and are frequently without access to credit services and adequate capital assets or not able to own large herds to manage households' daily requirements. Similar observations have been made by Kakota et al. ([2011]) in Malawi and Tesso et al. ([2012]) in Ethiopia that widowed or divorced household heads are more vulnerable because they rely on income earned by either the father or mother as the bread winners. These findings make a strong case for continuous targeting of pastoralist women in resilience-building interventions in the rangelands.

The results are consistent with previous findings (Deressa et al. [2008]) in similar ecosystems. However, for the biophysical variables, the greater the level of household reliance on natural resources, such as pastoralism or dry land crop farming, the greater will be their vulnerability to climate variability and change. This is partly because the use of such natural resources is dependent on rainfall, which is projected to change. This study observed that almost all the postulated biophysical/environmental variables contribute positively to household vulnerability. It is likely that the level of dependence on natural resources especially pastures and water will vary from household to household. For example, while the majority of households (78.9%) depend on livestock herding as their main source of livelihood, for others, livestock is just an equal or lesser contributor besides other economic activities.

The level of vulnerability is closely associated with the degree of poverty in Turkana. This county has higher degrees of vulnerability in terms of households' adaptive capacity or social and economic characteristics that are linked to their exposure and sensitivity to biophysical variables (Lo´pez-Carr et al. [2014]). The findings of this study suggest that in case of an external stress or shock, the majority of the households would need some assistance for them to recover. These results reflect the findings of studies by Kenya National Bureau of Statistics (KNBS) ([2013]) which showed that Turkana is the poorest county in Kenya, with 87.5% of the population living in absolute poverty. Other studies have indicated that Turkana region has a number of households with high of dependents, low engagement in off-farm activities and low levels of education (Blench[2000]; McPeak and Barrett [2001]; Watson and van Binsbergen [2008]).

The determinants of households' vulnerability were found to be significantly influenced by the sex of the household head, age of the household head, size of the household, number of dependents, marital status, social linkages, access to extension services and early warning information. In addition, non-farm income, herd size and diversity, herd structure and herd mobility, access to markets, households' employment status, coping strategies and access to credit were also observed to be the key determinants of the households' vulnerability to climate-induced stresses. This concurs with studies by Eriksen et al. ([2005]) and Notenbaert et al. ([2013]) which similarly observed some of these factors to be the key determinant of households' vulnerability to climate variability and change in rural communities. The results are also consistent with previous findings by Kakota et al. ([2011]) and Gebrehiwot and van der Veen ([2013]). From these findings, it seems there is still more to be done to understand vulnerability and its underlying processes. In this paper, the causal relationships between the statistically significant variables and outcomes in terms of vulnerability levels are explained using existing literature. Some of these explanations are, however, rather speculative and not confirmed for the local situation. More in-depth qualitative fieldwork, such as including open-ended questions, conducting focus-group discussions or in-depth interviews with selected households based on these factors, could strengthen our confidence in some of these explanations. The combination of quantitative surveys with qualitative autopsy is likely to provide a much more in-depth analysis of determinants of vulnerability.

CONCLUSIONS

Understanding vulnerability to environmental change and extreme climate events is necessary for policy makers to develop mitigation and adaptation programmes for long-term resilience. Vulnerability analyses contribute to the knowledge on climate-sensitive socio-economic or ecological systems, enabling policy to be targeted on the most vulnerable places, sectors or people and adaptation options to be defined. The results of this study from Turkana portray social and biophysical vulnerability indices which are useful to local development programming for long-term resilience. However, as a prerequisite for building households' resilience to climatic extremes, in-depth understanding is necessary of the adaptive capacity, exposure and sensitivity.

This study focused on the micro-level to assess the variability of vulnerability across different households. Categorization of vulnerability levels was used to help identify households that are not vulnerable currently but have a high probability of becoming vulnerable in the future. In the future, studies assessing household-level vulnerability to climate-induced stresses should explore the use of panel data as well as cross-sectional data to portray longitudinal and cross-sectional characteristics of households. Similarly, future resilience interventions should target individual households within a community because major decisions about adaptation to climate-induced stresses and livelihood processes are taken at the micro-level.

We conclude that because of various social, economic and biophysical determinants observed to influence households' vulnerability to climate-induced stresses, policies with emphasis on women's empowerment, promoting education, supporting extension services and enhancing diversifications of income sources and access to credit, supporting herd mobility and diversity, creating employments, and increasing access to markets and early warning information are likely to improve resilience of pastoral households. Although the results of this study are specific to Turkana County in Kenya, the approach and findings could be applicable to other arid and semi-arid areas in the region.

AUTHORS' CONTRIBUTIONS

FEOO collected the data and drafted the manuscript. OVW and MN guided the design of the research study and provided technical reviews of the paper. All authors read and approved the final manuscript.

ACKNOWLEDGEMENTS

The work has been supported by small grants from the African Climate Change Fellowship Program, Global Change for System Analysis, Research & Training (ACCFP/START), and additional support from the Red Cross/Red Crescent Climate Centre/START and United Nations International Strategy for Disaster Reduction (UNISDR) under Climate and Development Knowledge Network small grant. The authors also thank Boniface Makone of the Welcome Trust for the statistical guidance during the preparation of this paper.

REFERENCES

1. Adger, WN, and PM Kelly. 1999. Social vulnerability to climate change and the architecture of entitlements. Mitigation and Adaptation Strategies for Global Change 4: 253–266.
2. Adger, WN. 1999. Social vulnerability to climate change and extremes in coastal Vietnam. World Development 27: 249–269.
3. Adger, WN. 2006. Vulnerability. Global Environmental Change 16 (3): 68–281.
4. Blench, R. 2000. Extensive pastoral livestock systems: Issues and options for the future. GCP/JPN/005/ JPN. FAO (Food and Agriculture Organization of the United Nations), Rome, Italy.
5. Brooks, N, WN Adger, and PM Kelly. 2005. The determinants of vulnerability and adaptive capacity at the national level and the implications for adaptation. Global Environmental Change 15 (2): 151–163.
6. Callaway, JM. 2004. Adaptation benefits and costs:how important are they in the global policy picture and how can we estimate them? Global Environmental Change 14: 273–284.
7. Carpenter, S, B Walker, JM Anderies, and N Abel. 2001. From metaphor to measurement: resilience of what to what? Ecosystems 4 (8): 765–781.
8. Deressa, T, RM Hassan, and C Ringler. 2008. Measuring Ethiopian farmers' vulnerability to climate change across regional states, IFPRI discussion paper no. 806.
9. Eriksen, S, and K O'Brien. 2007. Vulnerability, poverty and the need for sustainable adaptation measures. Climate Policy 7: 337–352.
10. Eriksen, S, K Brown, and K Mick. 2005. The dynamics of vulnerability:locating coping strategies in Kenya and Tanzania. Geographical Journal 171: 287–305.
11. Fraser, EDG, AJ Dougill, K Hubacek, CH Quinn, J Sendzimir, and M Termansen. 2011. Assessing vulnerability to climate change in dryland livelihood systems:Conceptual challenges and interdisciplinary solutions. Ecology and Society 16 (3): 14.

12. Füssel, H. 2007. Vulnerability:a generally applicable conceptual framework for CC research. Global Environmental Change 17: 155–167.
13. Gallopın, GC. 2006. Linkages between vulnerability, resilience, and adaptive capacity. Global Environmental Change 16: 293–303.
14. Gebrehiwot, T, and A van der Veen. 2013. Climate change vulnerability in Ethiopia:disaggregation of Tigray Region. Journal of Eastern African Studies 7 (4): 607–629.
15. Greene, WH. 1997. Econometric analysis. Prentice Hall, New Jersey.
16. Hendrickson, D, J Armon, and R Mearns. 1998. The changing nature of conflict and famine vulnerability: the case of livestock raiding in Turkana District, Kenya. Disasters 22: 185–199.
17. Herrero, M, C Ringler, J van de Steeg, P Thornton, T Zhu, E Bryan, A Omolo, J Koo, and A Notenbaert. 2010. Climate variability and climate change and their impacts on Kenya's agriculture sector. International Livestock Research Institute (ILRI), Nairobi, Kenya.
18. Hewitt, K. 1995. Excluded perspectives in the social construction of disaster. International Journal of Mass Emergencies and Disasters 13: 317–319.
19. Hinkel, J. 2011. Indicators of vulnerability and adaptive capacity: towards a clarification of the science–policy interface. Global Environmental Change 21: 198–208.
20. 2012. Livelihood profiles:Six livelihood zones in Turkana County, Kenya. June 2012 report prepared for Oxfam and Save the Children UK.
21. 2001. Climate change 2001:Impacts, adaptations and vulnerability. Report of working group II. Synthesis report. Cambridge University Press, Cambridge.
22. 2012. Managing the risks of extreme events and disasters to advance climate change adaptation. A special report of Working Groups I and II of the Intergovernmental Panel on Climate Change. Cambridge University Press, Cambridge, UK.
23. 2014. Climate change: Impacts, adaptation, and vulnerability. Working Group II contribution to the IPCC 5th assessment report.

24. Jaetzold, R, and H Schmidt. 1983. Farm management in Kenya, vol. II part C: East Kenya. Ministry of Agriculture, Nairobi, Kenya.
25. Kabubo-Mariara, J. 2009. Global warming and livestock husbandry in Kenya:impacts and adaptations. Ecological Economics 68: 1915–1924.
26. Kakota, T, D Nyariki, D Mkwambisi, and W Kogi-Makau. 2011. Gender vulnerability to climate variability and household food insecurity. Climate and Development 3 (4): 298–309.
27. Kasperson, J, R Kasperson, and B Turner. 1995. Regions at Risk: Comparisons of Threatened Environments. United Nations University Press, New York.
28. Kelly, PM, and WN Adger. 2000. Theory and practice in assessing vulnerability to climate change and facilitation adaptation. Climatic Change 47 (4): 925–1352.
29. 2013. Exploring Kenya's inequality: Pulling apart or pooling together. National Bureau of Statistics and Society for International Development (SID), Nairobi, Kenya.
30. Klein, RJ, S Eriksen, LO Naess, A Hammill, TM Tanner, C Robledo, and K O'Brien. 2007. Portfolio screening to support the mainstreaming of adaptation to climate change into development assistance, Tyndall Centre Working Paper No. 102.
31. Liverman, DM, et al.. 1990. Vulnerability to global environmental change. In Understanding global environmental change, ed. Kasperson RE, 27–44. Center for Technology, Environment, and Development, Clark University, Worcester.
32. Lo´pez-Carr, D, GP Narcisa, EA Juliann, MM Jankowska, C Funk, G Husak, and J Michaelsen. 2014. A spatial analysis of population dynamics and climate change in Africa: potential vulnerability hot spots emerge where precipitation declines and demographic pressures coincide. Population Environment 35: 323–339.
33. Madu, IA. 2012. Partial vulnerability of rural households to climate change in Nigeria:Implication for internal security. Working paper number 2; May 2012. The Robert S. Strauss Center for International Security and Law, University of Texas, USA.
34. Marshall, NA, CJ Stokes, NP Webb, PA Marshall, and AJ Lankester. 2014. Social vulnerability to climate change in primary producers: a typology approach. Agriculture, Ecosystems and Environment 186: 86–93.

35. McCabe, JT, R Dyson-Hudson, PW Leslie, PH Fry, N Dyson-Hudson, and J Wienpahl. 1988. Movement and migration as pastoral responses to limited and unpredictable resources. In Arid lands today and tomorrow, ed. Whitehead EE, Hutchinson CF, Timmerman BN, Varady RG, 727–734. Proceedings of an International Research and Development Conference, Tucson, Arizona, USA.
36. 2001. Climate change 2001: Impacts, adaptation and vulnerability. Cambridge University Press, Cambridge.
37. McPeak, JG, and CB Barrett. 2001. Differential risk exposure and stochastic poverty traps among east African pastoralists. American Journal of Agricultural Economics 83 (3): 674–679.
38. Nabikolo, D, B Bashaasha, MN Mangheni, and JGM Majaliwa. 2012. Determinants of climate change adaptation among male and female headed farm households in eastern Uganda. African Crop Science Journal 20 (2): 203–212.
39. 2013. Turkana County drought monitoring and early warning bulletin - December 2013.
40. Nicholson, SE. 2014. A detailed look at the recent drought situation in the Greater Horn of Africa. Journal of Arid Environments 104: 71–79.
41. Notenbaert, A, SN Karanja, M Herrero, M Felisberto, and S Moyo. 2013. Derivation of a household-level vulnerability index for empirically testing measures of adaptive capacity and vulnerability. Regional Environmental Change 13: 459–470.
42. Opiyo, FEO, SM Mureithi, and RK Ngugi. 2011. The influence of water availability on pastoralist's resource use in Kitui and Mwingi districts of Kenya. Journal of Human Ecology 35 (1): 43–52.
43. Paavola, J. 2008. Livelihoods, vulnerability and adaptation to climate change in Morogoro, Tanzania. Environmental Science and Policy 11 (7): 642–654.
44. Pearson, L, R Nelson, S Crimp, and J Langridge. 2008. Climate change vulnerability assessment: Review of agricultural productivity. CSIRO climate adaptation flagship working paper 08.1.
45. Pulwarty, RS, and WE Riebsame. 1997. The political ecology of vulnerability to hurricane-related hazards. In Hurricanes:

Climate and socioeconomic impacts, 185–214. Springer, Berlin, Heidelberg, New York.

46. 2012. National policy for sustainable development of Northern Kenya and other arid lands; Sessional Paper No. 8 of 2012, Republic of Kenya.

47. Schilling, J, FEO Opiyo, and J Scheffran. 2012. Raiding pastoral livelihoods: Motives and effects of violent conflict in north-western Kenya. Pastoralism: Research, Policy and Practice 2: 25.

48. Sherwood, A. 2013. Community adaptation to climate change: exploring drought and poverty traps in Gituamba location, Kenya. Journal of Natural Resources Policy Research 5 (2–3): 147–161.

49. Silvestri, S, E Bryan, C Ringler, M Herrero, and B Okoba. 2012. Climate change perception and adaptation of agro-pastoral communities in Kenya. Regional Environmental Change 12 (4): 791–802.

50. Skjeflo, S. 2013. Measuring household vulnerability to climate change—Why markets matter? Global Environmental Change 23 (6): 1694–1701.

51. Stringer, LC, MS Reed, AJ Dougill, and C Twyman. 2009. Local adaptations to climate change, drought and desertification: Insights to enhance policy in southern Africa. Environmental Science and Policy 12: 748–765.

52. Tesso, G, B Emana, and M Ketema. 2012. Analysis of vulnerability and resilience to climate change induced shocks in North Shewa, Ethiopia. Agricultural Sciences 3: 871–888.

53. Thomas, RJ. 2008. Opportunities to reduce the vulnerability of dryland farmers in Central and West Asia and North Africa to climate change. Agriculture, Ecosystems and Environment 126 (1–2): 36–45.

54. Thornton, PK, PG Jones, T Owiyo, RL Kruska, M Herrero, P Kristjanson, A Notenbaert, N Bekele, and A Omolo. 2006. Mapping Climate Vulnerability and Poverty in Africa. Report to the Department for International Development, International livestock research institute, Nairobi, Kenya.

55. Turner, BL, RE Kasperson, PA Matson, JJ McCarthy, RW Corell, L Christensen, N Eckley, JX Kasperson, A Luers, ML Martello, C Polsky, A Pulsipher, and A Schiller. 2003. A framework for vulnerability analysis in sustainability science. Proceedings of the

National Academy of Sciences of the United States of America 100 (14): 8074–8079.

56. 2009. Global assessment report on disaster risk reduction.
57. 2004. Living with risk: A global review of disaster reduction initiatives. United Nations International Strategy for Disaster Reduction, Geneva, Switzerland.
58. 2011. Horn of Africa Crisis: Situation Report No. 13: United Nations Office for Co-ordination of Humanitarian Affairs.
59. Watson, DJ, and J van Binsbergen. 2008. Livestock market access and opportunities in Turkana, Kenya. ILRI Research Report 3, 106. ILRI (International Livestock Research Institute), Nairobi, Kenya.
60. Yohe, G, and RSJ Tol. 2002. Indicators for social and economic coping capacity—moving toward a working definition of adaptive capacity. Global Environmental Change 12: 25–40.
61. Yuga, NG, PG Shivakoti, and RP Sylvain. 2010. Household-level vulnerability to drought in hill agriculture of Nepal: Implications for adaptation planning. International Journal of Sustainable Development and World Ecology 17 (3): 225–230.

Chapter 7

Institutions for Managing Common-Pool Resources: the Case of Community-Based Shrimp Aquaculture in Northwestern Sri Lanka

Eranga K Galappaththi and Fikret Berkes

Natural Resources Institute, University of Manitoba, 70 Dysart Road, Manitoba, Winnipeg, Canada R3T 2N2

ABSTRACT

Large-scale shrimp aquaculture can have major social and environmental impacts. Can community-based approaches be used instead? We examined three coastal community-based shrimp aquaculture operations in northwestern Sri Lanka using a case study approach. These shrimp farms were individually owned by small producers and managed under community-level rules. The system was characterized by three layers of institutions: community-level shrimp farmers' associations; zone-level associations; and a national-level shrimp farming sector association. The national level was represented by a joint body of government and sector association. We evaluated the effectiveness of this institutional structure especially with regard to the management of shrimp disease that can spread through the use of a common water body. Lower operational costs make them a highly attractive alternative to large-scale aquaculture. In many ways, private shrimp aquaculture ownership with community-level institutions, and government supervision and coordination seem to work well.

INTRODUCTION

Can shrimp aquaculture be carried out through community-based institutions? And if so, can community-based aquaculture be considered a potential alternative to large-scale, company-owned aquaculture? Such questions have theoretical significance for commons management, and practical significance for sustainable aquaculture. We explore these questions based on coastal shrimp farming communities on the Indian Ocean in the northwestern part of Sri Lanka. The most common shrimp species cultured in these communities is the giant tiger prawn or Asian tiger shrimp (*Penaeus monodon),* a prime export species. Sri Lanka is not a big player in the global shrimp trade, and the operations in question have received little attention even in Sri Lanka itself (Galappaththi [2010]). In terms of production volume and technology, Sri Lankan shrimp aquaculture sector is relatively small compared to other Asian countries (MFARD [2011]; Anon.[2007]).

In most of South and Southeast Asia, shrimp aquaculture has a mixed history of social (Nayak and Berkes [2010]) and environmental impacts (Bhatta and Bhat [1998]; Primavera [1997]; Huitric et al. [2002]). Commercial operations are typically large-scale and company-owned. They have resulted in the destruction of coastal vegetation, including mangroves, that act as a buffer to protect coastlines from such disasters as tsunamis (Adger et al. [2005]). Historically, shrimp aquaculture operations have often collapsed after a few years of profit-taking, often the victim of shrimp diseases as a consequence of concentrating such large numbers of one species. For example, Huitric et al. ([2002]) tracked the rise and fall of large-scale shrimp farming in Thailand and around the Gulf of Thailand, leaving behind its wake collapsed farms and damaged land and waterscapes. However, in some areas, the collapse of large-scale farms creates the opportunity for the start of small, locally owned operations at the abandoned locations.

Such is the case in the northwestern part of Sri Lanka, where the country's large-scale shrimp farms used to be located and where the small-scale operations now exist. Sri Lanka's shrimp production reached a peak in 1998 (Munasinghe et al. [2010]), and then declined after a major outbreak of two virus diseases. In the place of large farms, numerous small-scale operations appeared over time, mostly family-level farms run by producers who had learned their trade by working earlier for the large companies. These producers patterned the new farms along the lines of agricultural operations, to take advantage of economies of scale afforded by community-based associations. These were associations that helped coordinate production, reduce pollution loads in the shared water resource (the key common-pool resource in this system), and functioned as a conduit of information between the farmers and the state.

Community associations have been known to play a major role in managing small-scale fisheries throughout the world (Digby [1975]; McCay et al. [2011]). Although there is a growing literature on such associations in finfish aquaculture (Haque et al. [2011]; Von Essn et al. [2013]), there is very little literature on their role in shrimp aquaculture. Even within Sri Lanka, the small-scale operations that replaced the collapsed industrial shrimp aquaculture are little known, much less appreciated (U. Amarasinghe, pers. comm.). How does small-scale aquaculture work? How can individual ownership be combined with community level coordination? As Bush et al. ([2010]) pose the question, how can small-scale shrimp farming techniques be integrated into a landscape level approach to protect mangroves and control water-borne shrimp diseases?

In the context of aquaculture, the commons problem occurs due to overuse of environmental resources as inputs and the addition of pollutants to the environment as waste (Pa´ez-Osuna[2001]). Solving the problem should start by addressing the two main issues related to commons management, the excludability problem and the subtractability problem (Ostrom et al. [1999]). Based on these two characteristics, Ostrom et al. ([1999]) define common pool resources as a class of resources for which exclusion is difficult and joint use involves subtractability. Controlling access to resources (the excludability problem) is addressed by the fact that the farms are privately owned. However, the private property regime is not ideally suited to deal with the subtractability problem when individual holdings are interconnected. Hence, rules are needed to reduce the impact of users on one another (Feeny et al. [1990]). Where community institutions exist, common property regimes may be suitable for users as an arena in which to make and enforce their own rules. Such community-based management works in many coastal resource contexts in Asia and elsewhere (Pomeroy and Rivera-Guieb [2006]; Berkes et al. [2001]). It is based on the idea that local resource users have a greater interest in the sustainable use of local resources than do large companies and other non-local players (Ostrom [1990]).

For common pool resources, collective management can work better than pure government management or private management (Ostrom et al. [1999]; Ostrom [2005]). Where users are interdependent, collective action is more appropriate than individual action and more effective for problem solving (Ostrom [2005]). This is where associations such as cooperatives come in. The International Cooperative Alliance defines cooperative as "an autonomous association of persons united voluntarily to meet their common economic, social, and cultural needs and aspirations through a jointly-owned and democratically-controlled enterprise" (ICA [2014]; Amarasinghe and Bavinck [2011]: 389). They work best in societies which have traditions of collective management (Yamamoto [1995]; Makino and Matsuda [2005]; Amarasinghe and De Silva [2001]). A cooperative functions as a local institution that can coordinate the response to the exclusion problem of commons, limiting access rights to the resource. It also deals with the subtractability problem, helping make and enforce local rules among the parties that have access rights, coordinating with government agencies and market institutions, and acting as an information broker (Beem [2007]).

Sri Lanka has a history of managing resources collectively in many sectors, including fisheries, aquaculture and agriculture (Amarasinghe and Bavinck [2011]; Amarasinghe [2006]). Historically, the Sri Lankan government has recognized and promoted "bottom-up" management through legislation aimed at developing and uplifting certain economic sectors (Anon. [2007]). Many of these resources are managed by cooperatives that facilitate collective decision making (Degen[1998]; Samarakoon et al. [2011]). Even though not all associations discussed in this paper are formally constituted as cooperatives, they function like cooperatives, despite varying designations.

However, the Sri Lankan literature on collective management does not include shrimp aquaculture carried out by community associations and cooperatives. The paper addresses this gap by, first, attempting to explore and understand how shrimp aquaculture is carried out by community-based institutions in northwestern Sri Lanka. Second, the paper discusses the evolution of small-scale shrimp aquaculture governance and the role of existing commons institutions. Finally, the paper explores the importance of community-based, small-scale aquaculture as a potential alternative to large-scale aquaculture.

METHODS AND STUDY AREA

This paper is based on a qualitative research study conducted during 120 days of fieldwork in three coastal communities of northwestern Sri Lanka (Figure 1). The communities were*Ambakandawila*, *Koththanthive*, and *Karamba,* hereafter referred to as communities A, B, and C respectively. These three communities were selected to capture the entire process of shrimp farming operation within the available limited timeframe. During the field work period, the three communities were in different stages of the shrimp farming process. For example, during the month of May, community A was in the middle of farming; community B was in the harvesting stage, and community C was in the pond preparation stage. The management activities and focus of the community-level shrimp farmers' associations varied according to the production stages of farmers in the community. For example, the main concern of community C was to finish postlarvae (shrimp "seed") stocking prior to the deadline. Community A— which was in the middle of the shrimp farming stage—focused on protecting their farms from shrimp diseases. Community B was in the process of shrimp harvesting and preparing ponds for the next culture season.

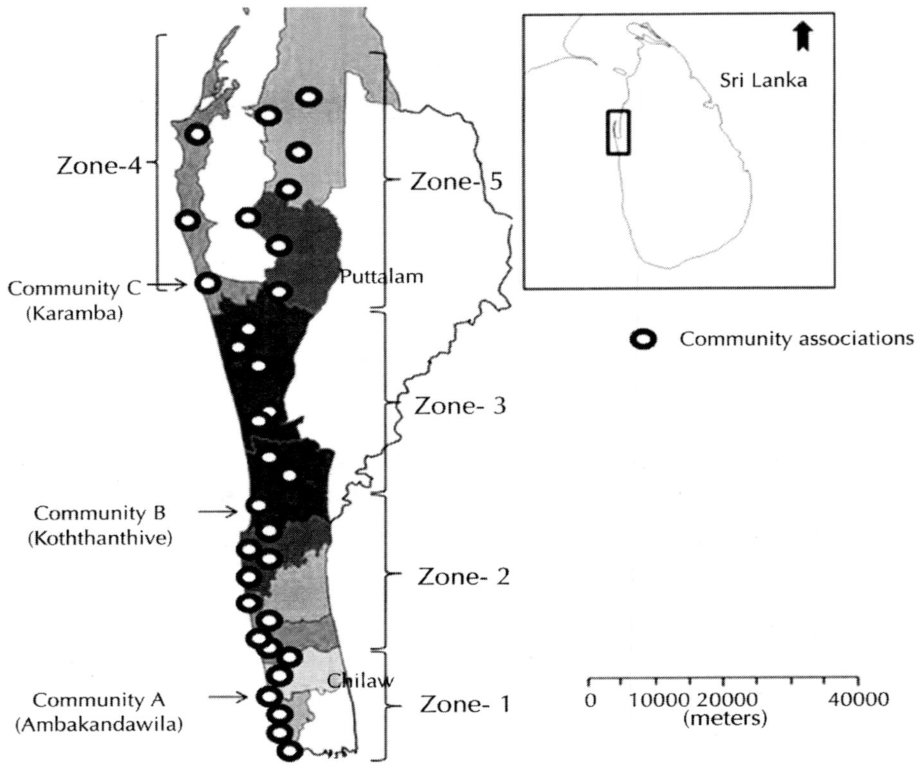

Figure 1: Study area: Three coastal communities (A, B, and C) in the north-western Sri Lanka and the distribution of communal institutions (Community associations/*Samithies*).

The research strategy adopted was the case study approach (Yin [2009]). Primary data were collected through: a) participant observation; b) semi-directive interviews with shrimp farmers and the leaders of shrimp farming community associations; c) focus group discussions; and d) key informant interviews with influential people involved in shrimp farming. Data collection took place from April to August in 2012. The sample of farmers was selected using a snowball sampling technique. Thirty eight (38) shrimp farmers and community associations (A: 13; B: 11; and C: 14) were interviewed. Further, the data were validated through seven key informant interviews and three focus group discussions. The data from participant observations helped to obtain a better contextual understanding of the shrimp farming areas. During the field work, a couple of community associations, which did not belong to the study sample, also invited the researcher to visit and observe their meetings.

Data gathered through participant observations and semi-directive interviews were documented immediately in the field. Analysis of findings began with the labeling and coding of data to help to distinguish differences and similarities. Mind maps and diagrams on concepts and, management processes, were developed based on the contextual understanding developed during field work. Insights and experiences gained through participant observation also helped the researcher to simplify the analysis process. Descriptive statistics (e.g., frequencies, percentages, average values) were used in assessing the background of community-based shrimp aquaculture in northwestern Sri Lanka. Microsoft Excel was used for data processing.

Community A is located in the southern part of northwestern coastal area, close to *Chilaw* lagoon. This is a rural community with people of Sinhalese origin. At the time of research there were about 150 households involved in small motorized craft fishing, shrimp breeding (hatchery), and shrimp feed selling in addition to shrimp farming. Community B is located in the middle of northwestern coastal area on the *Mundel* lagoon. This is also a rural isolated community but with Tamil ethnic backgrounds. The size of community B is about 100 households. The main household income generating activities are: fishing; paddy farming; and government employment. Community C is located in the northern part of the northwestern coastal area, close to the *Puttalam* lagoon. This community has a mix of people with Sinhalese, Tamil, as well as Muslim origins. Salt extraction and coconut cultivation were the main household activities, other than shrimp farming.

DESCRIPTION OF THE AQUACULTURE OPERATIONS

Almost all of the existing shrimp aquaculture operations in northwestern Sri Lanka are small-scale. The number of ponds operated in a farm can be considered as an indicator of the extent of shrimp farming. In 2011–12, 37 percent of the sampled farmers operated two to five ponds, whereas 11 percent operated just one pond. Twenty-nine percent and 18 percent of shrimp farmers operated five to ten and ten to fifteen ponds, respectively. Only five percent of farmers had more than 15 ponds. The size of the smallest pond was about 0.2 hectares whereas the largest was about 0.8 hectares. As of 2013, there were many abandoned shrimp ponds in the northwestern coastal area, and only about 45 percent of ponds is operational. Small-scale shrimp farmers do not invest money in the construction of new ponds; rather they use existing abandoned farms to reduce costs. Another indicator of the scale is the total land area of the farm (i.e., the full extent or the size of the farm), including the ponds which were being operated and rested. Fifty five percent of shrimp farmers had farming areas between one and three hectares in size. Thirteen percent of farmers had less than one hectare of land. Sixteen percent and 11 percent of farmers had farm areas between three and five and five and seven hectares, respectively. The recorded percentage of farmers with more than seven hectares of farm lands was five percent. Only the farmers from community C used more than seven hectares of farm land. Multiple reasons were given by the farmers for resting some of the ponds in their farms: lack of labor; relatively low profit margins and increasing cost; limited shrimp stocking capacity issued by the farmers' co-operative; and pond disinfecting to reduce disease risk. Draining a pond and exposing it to sunlight presumably reduces disease risk.

Shrimp farms were owned and operated by individuals, not by community associations or large private corporations. However, 53 percent of these farms were on government-owned lands without any compensation paid for using the lands for commercial purposes. There were unresolved legal issues mostly related to land ownership, and it seemed political influences were in play. Thirty-seven percent of farmers used their own private lands. The remaining 11 percent of farmers rented lands from shrimp farmers who had discontinued farming previously. Due to the high input cost of farming, the general practice of farmers was to minimize the number of employees as a way to reduce costs. In most cases, family members provided the labor needed. Involvement of family members in the farming operation was common. Seventy-four percent of shrimp farmers said they receive the support and involvement of their family members and/or relatives. Sixty four percent of them received support from female family members --wives, daughters, mothers, and/or other close relatives. These women were involved in shrimp farming in diverse ways: labor (shrimp harvesting); book keeping; day-to-day management activities; financial contributions towards farming; and community association activities. Complementing the individual/family-level farm ownership, community level farming activities were governed by community-level rules designed and enforced by the community institutions as a part of a multi-layered management structure.

MULTI-LAYERED MANAGEMENT STRUCTURE

Shrimp aquaculture operations are managed under a commons regime consisting of three kinds of institutions: private; communal; and governmental. At the heart of this management regime is the shrimp farmers' (or community) association, called *samithiya* in Sinhala. The distribution of community level associations in the northwestern part of the island is shown in Figure 1. These institutions are self-organized and formulate and implement their own rules. For example, community-level rules specify that a pond should not be harvested when shrimp are less than one month old or less than five grams in average weight (community A); water should not be released or pumped until the two month culture period is completed (community A; community B); and partial harvests are not allowed (community A; community C). Obtaining community association membership is a requirement for any person starting-up a shrimp farm. Community associations have collective authority and a high level of involvement in regulating and managing shrimp aquaculture. Elected officers of the community associations enforce the rules and assess penalties for those not conforming. Penalties vary from fines to denial of permit—a document issued by the community association required for carrying out shrimp farming during a particular crop season.

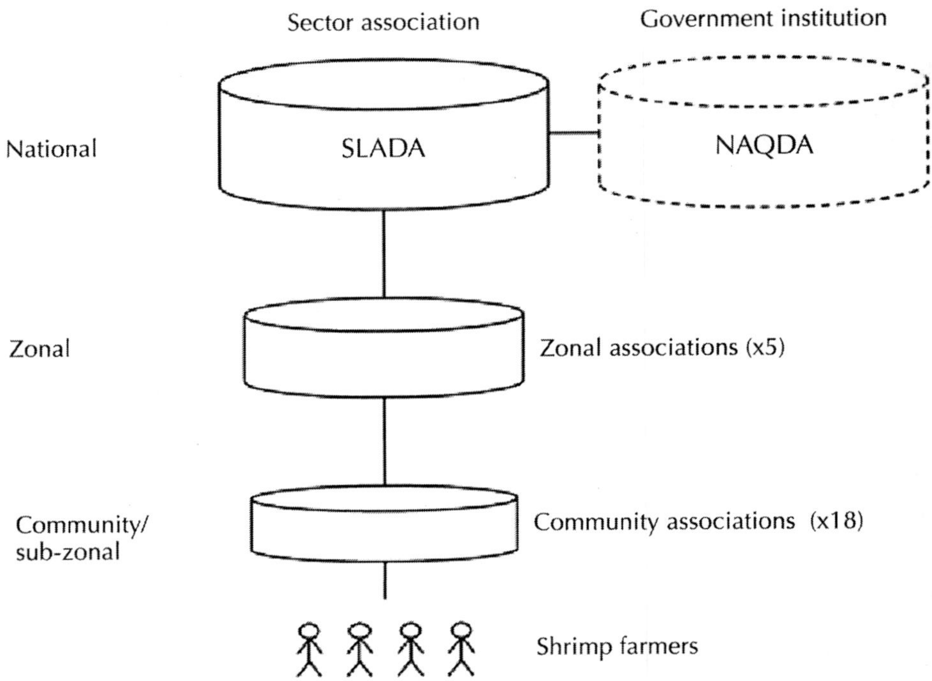

Figure 2: Evolution of small-scale shrimp aquaculture.

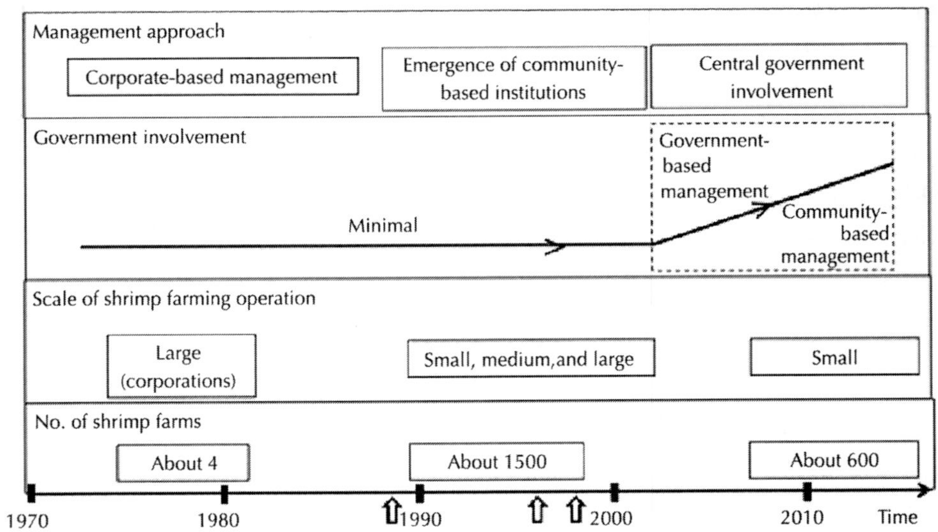

Figure 3: Multi-level structure of commons institutions.

Most of the community associations consist of 20–60 members. Members pay an annual membership fee. The elected officials of a community association include: president, vice president, secretary, treasurer, and assistant treasurer. These officers have the authority to make decisions on behalf of the membership—a power granted through the collective agreement. However, the officials are also responsible for elaborating and adequately justifying such decisions during the upcoming meetings. The community associations collectively make decisions on the day-to-day regulatory operations related to community level shrimp aquaculture. For example, monitoring the application of proper management practices for shrimp disease prevention; inspection of ponds prior to issuing post larvae bill; and supporting members to obtain electricity connection to farms, are some of the major activities. Most of the associations have government aquaculture extension officers working closely with them to ensure that practices comply with national level regulations. However, the extension officers are merely observers and are not part of the decision-making. Collective decision making is one of the important characteristics of community associations. Moreover, community associations act as hubs for sharing information required for operations. These community institutions are inter-connected through zonal-level associations. Table 1 shows characteristics of the three community associations studied.

Table 1: Characteristics of community associations

Characteristics	Community A	Community B	Community C
Year of establishment	2006	1995	1995
Nature of origin (farmer-initiated or government-initiated)	Government-initiated	Farmer-initiated	Farmer-initiated
Registered year	2012	2002	1998
Registered under	NAQDA	Divisional secretariat office	Cooperative Act
Current membership	30	61	30
Number of officers	7	7	11

Availability of a constitution	Yes	Yes	Yes
Bank account/ Letter head/ Official rubber stamp	Yes	Yes	Yes
Meeting frequency	Monthly, special, and annual meetings	Monthly and special meetings	Monthly and special meetings
Governance structure	Mostly flat	Hierarchical	Flat
Cohesiveness	High	Moderate	Very high

Galappaththi and Berkes

One basic feature of shrimp aquaculture in the study area is the zonal system design introduced by the SLADA[a] (Sri Lanka Aquaculture Development Association) for the main purpose of managing disease. The shrimp farming area is divided into five main zones and 32 sub-zones. Some zones include just one sub-zone whereas the others represent up to eight sub-zones. If the number of farmers in a particular sub-zone is too small, then adjacent sub-zones are amalgamated to form a single association to represent a cluster of sub-zones. Leaders of sub-zonal associations represent the zonal association. All zonal and sub-zonal associations are collectively represented in SLADA with at least one representative from each zone and sub-zone.

SLADA is the national level association, which represents all the direct stakeholders in the shrimp farming sector—farmers; hatchery owners; feed suppliers; and shrimp processors. This association was established in 2005 with an initial membership of 16 people. Later, zonal and sub-zonal community associations were allowed to become members, resulting in a total membership of about 50 (about 20 institutional members). SLADA is an independent body. Even though the main objective of SLADA is to develop the shrimp aquaculture sector, initially the mandate was not restricted to shrimp aquaculture. Development and planning of ornamental fish aquaculture and inland aquaculture were also included in the mandate. In developing the shrimp farming sector, SLADA appointed six committees targeting different subject areas, such as environmental protection; hatchery development; and shrimp farm development. Moreover, SLADA and NAQDA (National Aquaculture Development Authority)—the government line institution responsible for aquaculture development and management, which falls under the of Ministry of Fisheries and Aquatic Resource Development in Sri Lanka —jointly developed a technical advisory committee representing expertise from local universities and research institutes. In 2006, SLADA was legally registered as a corporation by the members themselves (Figure 2).

EVOLUTION OF SMALL-SCALE SHRIMP AQUACULTURE GOVERNANCE

Evolution of small-scale shrimp aquaculture in northwestern Sri Lanka is related to the collapse of large-scale shrimp farming operations, the management approach, government's involvement and impact, and the management of the key common-pool resource, water. Figure 3 illustrates the evolution of shrimp aquaculture from the 1970s to 2012.

Collapse of Shrimp Farming Operations

In the late 1970s, shrimp aquaculture in Sri Lanka started off as a commercial-scale venture. Initially, the farms were geographically limited to the eastern part of the country (Batticaloa District) and only four companies (Carson, Enris, Lever Brothers and Andrew's) were in the industry. To begin with, all the operations were done in an integrated fashion, whereby shrimp breeding (hatcheries), farming, and processing, as well as exporting activities were managed by the same company. According to the respondents, some farms were as large as 150 ha in area with hundreds of ponds. At the time, these companies reportedly earned attractive profits. The information related to culture systems, input supply, and marketing were accessible only to a few stakeholders attached to these companies. During the early 1980s, the profit making potential of this new industry attracted investors such as politicians, bank managers, and wealthy business owners. At the same time, some of the employees of these pioneer companies started their own farms in eastern and northwestern provinces.

Later (in the 1990s), the government intervened by promoting shrimp farming as a profitable self-employment opportunity among locals, leading to many other community-level small-scale farmers entering the industry. As a result, farming area expanded over the northwestern part of the country. Small-scale farmers conducted their operations within their community areas, while large and medium-scale farmers shifted from place to place by converting mangrove forests (Gunawardena and Rowan [2005]) and coconut plantations into shrimp farms (Cattermoul and Devendra [2002]). However, the expansion of aquaculture into the eastern part of the country was restricted due to the civil war that was taking place. During this time, there were outbreaks of three viral diseases. The first major disease outbreak Monodon Baculo Virus (MBV) came in 1988/9. The second major disease outbreak was the White Spot Syndrome Virus (WSSV) in 1996, followed by a third outbreak in 1998 with both WSSV and Yellow Head Virus (YHV). Most of the pioneer companies shut down operations a result of these outbreaks.

As of 2012, almost all the 600 shrimp farms in the northwestern Sri Lanka are small-scale operations. Large or medium-scale shrimp operations have not survived under disease conditions. As documented earlier, the average size of a shrimp farm is small compared to the operations that took place at the beginning of the industry.

Changes in Management Approach

The type of management system found in the 1970s and 1980s was a corporate-based management with four large multi-national companies controlling the industry. However, the decline of corporate shrimp farming during the 1990s created space for small-scale operations with their own associations. This is the era of emergence of community-based institutions across northwestern Sri Lanka. Community associations were registered under different legislations/authorities such as the Cooperative Act, divisional secretariat offices, and most recently with government institutions like NAQDA.

Before 2003, there were no governmental, technical or other controls over shrimp farming activities. Resource management activities were mostly limited to self-managed community-level associations, even though these associations were not recognized by the government. There was neither any proper linkage between farmers and government nor proper management over shrimp aquaculture production and quality. For instance, there was no specific timeframe for stocking shrimp postlarvae. Shrimp farmers and hatcheries continued production throughout the year, and people were free to enter and leave shrimp farming whenever they wished.

Government's Involvement and Impact

Until 2003, government involvement in shrimp aquaculture was minimal and was mostly restricted to divisional and provincial level general administration, and research on water quality of the northwestern lagoons. The direct involvement of the central government in the shrimp aquaculture sector started in 2003/4 through NAQDA. This action led to many changes in management, resulting in a collaborative management approach. NAQDA's approach was to work in collaboration with SLADA and community-level shrimp farmer and breeder (hatchery owner) associations. Initially, it was not easy for the independent-minded actors in the sector (especially shrimp farmers and breeders) to adjust to the new management system, as they were skeptical of the purposes behind government involvement.

Design and introduction of the zonal crop calendar system in 2005/06 was one of the initial tasks of SLADA and the individual community institutions. The crop calendar system emerged out of a study conducted by a few of the founding members of SLADA with previous experience and knowledge of aquaculture. The study report recommended management that involved the development of a zonal crop calendar system. This crop calendar system was initiated by SLADA in collaboration with NAQDA. The calendar system was designed to limit farmers accessing common water bodies through a schedule or timetable leading to a major change in the way farming operations were conducted. Under this system, farmers were not allowed to access the water bodies at any time but only during a particular period when all of the community farmers of a subzone could access the same water body. The initial reaction of shrimp farmers and hatchery owners was to resist, mainly because the crop calendar limited the production cycle (number of crops per year). Moreover, the introduction of new standards and quality improvement measures targeting shrimp farms and hatcheries led to social conflicts between shrimp farmers/hatchery owners and NAQDA. Adoption of these measures required costly physical infrastructure changes such as construction of chlorine baths at farm and hatchery entrances to control disease spread. NAQDA field extension officers were tasked with monitoring these standards by visiting farms and hatcheries. However, these officers at times were not allowed to enter farms and hatcheries—and there were reported incidences of serious verbal conflicts and even physical harassment.

During the first few crop cycles under the new zonal crop calendar system, some relatively large-scale shrimp farmers and large-scale hatchery owners in Sri Lanka left the industry as it was not profitable to maintain their farms and hatcheries in an idle state during parts of the year as they still had to incur costs in terms of salaries and fixed costs such as maintenance. The overall effect of these changes was a decrease in national shrimp aquaculture production after 2003. However, farmers and hatchery owners who remained in the sector experienced a very successful crop year after the introduction of the zonal crop calendar. Farmers had a good harvest due to reduction in disease (fewer infected ponds), and hatchery owners received a good price for postlarvae. After realizing the success, some farmers who had left earlier, joined again in the following season; however, the next season was not as successful, in part due to lower market prices.

During the period, 2006 to 2009/10, there were ups and downs in terms of shrimp production. People started to think that the white spot disease had been eradicated, but it returned from time to time. Most of the medium-scale farmers also left farming due to the uncertain nature of production and significant increases in costs of labor, fuel, and chemicals such as chlorine. Although small-scale farms were affected by the same factors, small operations tended to have less operational and financial risks, mainly due to use of family labor.

By this time, the majority of farmers and hatchery owners had adapted to the new crop calendar system. NAQDA, with the support of SLADA, further strengthened the existing community associations while establishing new associations in certain other communities. NAQDA's aquaculture field extension officers were able to visit farming communities without any problems, and NAQDA itself became more flexible after their experience. Some illegal activities such as shrimp postlarvae stocking and releasing disease-infected water into the natural water body, still occurred, but NAQDA and SLADA are able to take action against these activities with the support of community-level associations. Shrimp aquaculture management system became well established as a result of these changes.

Based on the SLADA report recommending a crop calendar system, different tasks were assigned to different levels of management. Community (primary) associations regulate community-level farming practices with the support of NAQDA field extension officers. For this purpose, there is a shrimp farm monitoring and extension unit in *Battulu Oya* established in 2008. It is the responsibility of second-level zonal associations to develop the infrastructure of shrimp farming areas in collaboration with divisional secretariat offices and provincial councils. Maintaining links (mainly communication) between the different management levels is also the role of zonal associations. Finally, the role of top-level management (SLADA and NAQDA) is to develop an effective national-level shrimp aquaculture system.

Water: Managing the Key Common-Pool Resource

Shrimp farms take water from a common source at the beginning of the production cycle and discharge used water at the end of the cycle, after shrimp have been harvested. Almost all the shrimp farms in the northwestern area are directly or indirectly connected to single water bodies, mainly via a canal that connects them (the Dutch canal) and the lagoons. *Puttalam* lagoon (28,000 ha), *Mundal* lagoon (3,600 ha), and *Chilaw* lagoon (700 ha) are the three main lagoons connected by the Dutch canal. *Kala Oya, Mi Oya, Daduru Oya,* and *Maha Oya* are the four main rivers connected to this system. The interconnected nature of this common water body makes the shrimp farms susceptible to shrimp diseases. Hence, one of the major tasks of the aquaculture management system is to control the spread of disease. Wetland ecosystem biodiversity and ecosystem services such as water filtration through mangroves may also have a role to play in this regard but their importance has not been investigated.

Regarding the common water supply as the key resource, the excludability problem (Ostrom et al.[1999]) is addressed by community associations with their own rules to exclude outsiders from shrimp farming in their communities. This is accomplished in two ways. First, the community association organizes the membership and oversees how and when the members access the water source at the given time period, as specified by the zonal crop calendar (main aim is to control the spread of any diseases). Second, if an outsider wants to start shrimp farming in a particular community, attending association meetings for a minimum period of six months is mandatory to be eligible for membership. This rule is in place to discourage community dwellers/ outsiders, who intend to start shrimp farming, from moving from one farming location to another within a short period of time, i.e. relatively irresponsible shrimp farming. This rule also ensures that the prospective members understand how shrimp farming operations are conducted within the community in a collective manner. The subtractability problem is associated with the release of waste water into the common water body. If a particular pond is disease infected and if the farmer releases contaminated pond water to the common water body, there is a high probability for disease spread into other farms. Disease cannot be completely eradicated but it can be controlled. Constant vigilance and monitoring by farmer families themselves make early detection possible to deal with shrimp disease. If an outbreak is caught early, shrimp can still be harvested quickly and sold.

Large-scale shrimp disease problems, such as those that led to the demise of the aquaculture companies in the 1990s, mainly occur when uncontrolled amounts of disease-infected water are released into a common water body. By limiting the amounts of waste water and staggering the times of release, the zonal crop calendar system reduces the incidence of disease, and helps farmers to control infections. Compliance is monitored and managed by community level associations making use of their own rules. These rules are implemented by the collectively elected officers of community associations, and enforced by higher level authorities as appropriate. According to Gazette number 1536/12-(February 13, 2008), shrimp can only be farmed in compliance with the crop calendar. To operate a shrimp farm, farmers must obtain a permit from NAQDA (as specified in Gazette number 1677/7-October 25, 2010) issued under the consent of the community association. Punishments for free riders have been specified depending on the severity of the violation and range from fines paid to the community association to court cases filed by NAQDA.

DISCUSSION AND CONCLUSIONS

Most of the commercially managed aquaculture operations in South and Southeast Asia are large-scale, corporate-based industries that often produce negative social and environmental impacts. Unsustainablity of large-scale aquaculture is well documented, but there have been some major attempts to make them more sustainable (Primavera [2006]; Bush et al. [2010]). Many large-scale aquaculture operations are known to move from one area to another, in order to avoid disease. Large-scale aquaculture tends to make its profit during the first few years, with the expectation that they can move onto another area, rather than aiming to establish sustainable practices and remain in one area (Huitric et al. [2002]).

By contrast, small-scale community-based aquaculture farmers do not have the option of relocating, and have strong incentives to seek ways to make sustainability work. They cannot afford to act in ways that will undermine their livelihoods, if they have attachment to place and intend to continue to live in that area as part of the local community. These considerations are well documented by commons scholars for a range of resource types in many parts of the world (Ostrom [1990]; [2005]). Thus, by design, small-scale aquaculture has built-in incentives to be sustainable (or else go out of business). Based on the Sri Lanka experience, the survival of small-scale aquaculture (as opposed to large-scale aquaculture which has not) seems to be related to a number of factors:

A mixed governance regime that combines private ownership with community-level commons institutions, which deal with exclusion and subtractability problems.

A management system that is able to limit and control the release of waste water, minimizing its environmental impact and the transmission of shrimp disease. Small-scale operators have low overheads and low operating costs, as family members work in the farm.

A close relationship with aquaculture ponds, allowing for constant monitoring by farmer families themselves for shrimp disease. There are relatively small net losses even under disease conditions, since disease detection is fast and the potential for limiting economic damage is relatively high. This is because the shrimp can still be sold at market prices, if disease is detected at the initial stage.

The social enterprise nature of small-scale aquaculture, whereby operations are based on an economic model in which resources provide for broader goals, such as improving human and environmental well-being, rather than merely maximizing profits for external shareholders (Ridley-Duff and Bull [2011]).

Competition over scarce resources is a common phenomenon throughout the world, and interactions among users tend to be competitive. However, in the context of small-scale shrimp aquaculture in Sri Lanka, farmers co-operate to overcome common challenges, such as disease. Such cooperation is not common in the world of free enterprise, and as Bruno Dyck (pers. comm.) pointed out; it can be considered as a different way of doing things, even a different worldview. There seem to be two major factors that shape the operations of Sri Lanka's small-scale aquaculture producers.

First, shrimp farmers are empowered by their experience having worked for the old (and failed) aquaculture companies; they constantly apply this knowledge (stocking, growing and harvesting shrimp) to the practice of their small-scale operations. Second, starting at the local level, multi-level commons institutions have been developed by shrimp farmers in cooperation with the government, for commons management (Berkes and Davidson-Hunt [2008]). These commons institutions function collaboratively, using a participatory approach. There are no formal co-management arrangements with the government. However, small-scale aquaculture, as practiced in Sri Lanka, functions like co-management because it involves shared decision-making between government and producers. The overall system is a hybrid regime of private, communal, and state commons management, in which government devolves certain powers to other parties.

The evolution of this commons management model, which dates back to 1990s, has been relatively rapid and has taken place under certain historical conditions. The community-based shrimp aquaculture management system in northwestern Sri Lanka seems to have evolved to its current state under the influence of two major drivers. The first is the tendency of large aquaculture operations in Sri Lanka in the 1970s and 1980s to develop disease. Shrimp disease problem is a key factor in the demise of large operations in general (Bush et al. [2010]). Disease spreads through the interconnected water bodies, which tend to be controlled centrally in large-scale operations, but separately in small-scale operations through the zonal crop calendar system. The second driver is the ability of Sri Lankan producers to use collective approaches. There is a historical and cultural tendency in Sri Lanka towards a collective approach in managing and sharing resources (Gunawardena and Steele [2008]).

Backbone of the community-based shrimp aquaculture management system is multi-level commons institutions. These institutions have been developed by shrimp farmers themselves, in cooperation with the government. This bottom-up multi-level institutional structure is effective in managing small-scale shrimp aquaculture management. Small-scale shrimp aquaculture owners are families. Their operations are owner-managed under community-level institutions, with government supervision and coordination. This private-communal-state mixed regime appears to be a viable alternative to "big aquaculture".

The current (as of 2013) small-scale shrimp aquaculture systems in Sri Lanka are by no means perfect. Shrimp farmers deal with new problems on a daily basis and constantly argue among themselves in local associations. As new challenges appear, they debate new approaches. The present management system will likely continue to evolve in the face of numerous changes in the social-ecological system into the future. However, the recent historical experience with northwestern Sri Lanka shrimp aquaculture systems suggests four lessons: (1) Small-scale shrimp aquaculture operations have been able to survive in an environment in which large-scale ones have not; (2) this has been mainly due to the ability to control the spread of disease under the recent management regime for small-scale shrimp aquaculture; (3) the mixed regime (private-communal-state) commons management model with multi-level institutional linkages has been effective in providing private incentives, community-level rules and government coordination and oversight; and (4) continuous learning and adaptation, such as adjusting the crop calendar from year to year and combining farmer knowledge with government technical knowledge, has provided a pathway to economic sustainability.

The present study is significant for several reasons. Even though small-scale community-based shrimp aquaculture practices exist elsewhere in the world (for example in Thailand, Berkes, unpublished field notes), it seems that they have not been documented or studied. The present study attempts to shed light on this area by confirming the existence of community-based operations and by providing evidence that such operations seem to be viable. Further, the study suggests that the comparatively lower environmental and social impacts of these operations make them highly attractive as an alternative approach. Principles and practices of Sri Lanka's community-based management system provide lessons in multi-level governance that may be applicable to other aquaculture operations.

AUTHORS' CONTRIBUTIONS

FB drafted substantial portions of the introduction and discussion sections of the manuscript. He also revised some of the critical intellectual content and the content flow. Both authors read and approved the final manuscript.

ACKNOWLEDGEMENTS

We thank the shrimp farmers of northwestern Sri Lanka, and gratefully acknowledge Rohan Paaris from the Sri Lanka Aquaculture Development Association for his invaluable support and guidance extended during the field work period. We thank Upali Amarasinghe (University of Kelaniya, Sri Lanka), Bruno Dyck (University of Manitoba), and Iroshani Tennakoon (Manitoba Health) for help and for sharing valuable thoughts and experiences in this field.

REFERENCES

1. Adger, WN, TP Hughes, C Folke, SR Carpenter, and J Rockström. 2005. Social-ecological resilience to coastal disasters. Science 309: 1036–1039.
2. Amarasinghe, US. 2006. Culture-based fisheries development in Sri Lanka: a case study. In Better-practice approaches for culture-based fisheries development in Asia, ed. Silva SS, Amarasinghe US, Nguten TTT, 50–72. ACIAR Monograph 120, Canberra, Australia.

3. Amarasinghe, O, and M Bavinck. 2011. Building resilience: fisheries cooperatives in southern Sri Lanka. In Poverty mosaics: realities and prospects in small-scale fisheries, ed. Jentoft S, Eide A, 383–406. Springer, Dordrecht.
4. Amarasinghe, US, and SS De Silva. 2001. Sri Lankan reservoir fishery: a case for introduction of a co-management strategy. Fisheries Management and Ecology 6 (5): 387–400
5. 2007. Ten year development policy framework of the fisheries and aquatic resources sector (2007–2016), Ministry of fisheries and aquatic resources, March 2007, 4.
6. Beem, B. 2007. Co-management from the top? The roles of policy entrepreneurs and distributive conflict in developing co-management arrangements. Marine Policy 31: 54–549.
7. Berkes, F, and I Davidson-Hunt. 2008. Commons in a multi-level world. International Journal of the Commons 2 (1): 1–6.
8. Berkes, F, R Mahon, P McConney, R Pollnac, and R Pomeroy. 2001. Managing small-scale fisheries: alternative directions and methods. International Development Research Centre, Ottawa.
9. Bhatta, R, and M Bhat. 1998. Impacts of aquaculture on the management of estuaries in India. Environmental Conservation 25 (2): 109–121.
10. Bush, SR, PAM van Zwieten, L Visser, H Van Dijk, R Bosma, WF De Boer, and M Verdegem. 2010. Scenarios for resilient shrimp aquaculture in tropical coastal areas. Ecology and Society 15 (2): 15.
11. Cattermoul, N, and A Devendra. 2002. Effective management for biodiversity conservation in Sri Lankan coastal wetlands, A measurement of the ecological footprint of shrimp farming in the Chilaw lagoon area. University of Portsmouth Cemare, Portsmouth, UK.
12. Degen, P. 1998. Fisheries co-management in Sri Lanka: learning from the experiences of the others. Mekong Fisheries Network Newsletter 4 (1): 1–2.
13. Digby, M. 1975. Organisation of Fishermen's Cooperatives. Plunkett Foundation, Oxford, UK.
14. Feeny, D, F Berkes, BJ McCay, and JM Acheson. 1990. The tragedy of the commons: twenty-two years later. Human Ecology 18 (1): 1–19.

15. Galappaththi, EK. 2010. Exploratory study on supply chains of cultured prawns in Sri Lanka. Postgraduate Institute of Agriculture, University of Peradeniya, Sri Lanka.
16. Gunawardena, W, and JS Rowan. 2005. Economic valuation of a mangrove ecosystem threatened by shrimp aquaculture in Sri Lanka. Environmental Management 36 (4): 535–550.
17. Gunawardena, A, and P Steele. 2008. The stake-net fishery association of Negombo lagoon, Sri Lanka: why has it survived over 250 years and will it survive another 100 years? In Promise, trust and evolution: managing the commons of South Asia, ed. Ghate R, Jodha N, Mukhopadhyay P, 144–154. Oxford Scholarship, New York.
18. Haque, ABMM, LE Visser, and MM Dey. 2011. Institutional arrangements in seasonal floodplain management under community-based aquaculture in Bangladesh. Asian Journal of Agriculture and Development 8 (1): 1–18.
19. Huitric, M, C Folke, and N Kautsky. 2002. Development and government policies of the shrimp farming industry in Thailand in relation to mangrove ecosystems. Ecological Economics 40: 441–455.
20. 2014. International Co-operative Alliance.
21. Makino, M, and H Matsuda. 2005. Co-management in Japanese coastal fisheries: institutional features and transaction costs. Marine Policy 29: 441–450.
22. McCay, BJ, W Weisman, and C Creed. 2011. Coping with environmental change: Systematic responses and the roles of property and community in three fisheries. World Fisheries: A Social-Ecological Analysis, eds. Ommer, RE, RI Perry, K Cochrane, and P Cury. 381–400. Wiley-Blackwell, Oxford.
23. [MFARD. 2011. Ministry of Fisheries and Aquatic Resource Development of Sri Lanka. Sri Lanka: . . Accessed 14 May 2014.
24. Munasinghe, MN, C Stephen, P Abeynayake, and IS Abeygunawardena. 2010. Shrimp farming practices in the Puttallam district of Sri Lanka: implications for disease control, industry sustainability, and rural development. Veterinary Medicine International 2010: 1–6.

25. Nayak, PK, and F Berkes. 2010. Whose marginalization? Politics around environmental justice in India's Chilika lagoon. Local Environment 15 (6): 553–567
26. Ostrom, E. 1990. Governing the commons: the evolution of institutions for collective action (11–13). Cambridge University Press, New York.
27. Ostrom, E. 2005. Understanding institutional diversity. Princeton University Press, Princeton.
28. Ostrom, E, J Burger, CB Field, RB Norgaard, and D Policansky. 1999. Revisiting the commons: local lessons, global challenges. Science 284: 278–282.
29. Pa´ez-osuna, F. 2001. The environmental impact of shrimp aquaculture: causes, effects, and mitigating alternatives. Environmental Management 28 (1): 131–140.
30. Pomeroy, RS, and R Rivera-Guieb. 2006. Fishery co-management: a practical handbook. International Development Research Centre, Ottawa.
31. Primavera, JH. 1997. Socio-economic impacts of shrimp culture. Aquaculture Research 28: 815–827.
32. Primavera, JH. 2006. Overcoming the impacts of aquaculture on the coastal zone. Ocean and Coastal Management 49: 531–545.
33. Ridley-Duff, RJ, and M Bull. 2011. Understanding Social Enterprise: Theory and Practice. Sage Publications, London.
34. Samarakoon, JI, M Nightingale, R Hermes, BL Joseph, and V Salagrama. 2011. Review of Community-based Integrated Coastal Management: Best Practices and Lessons Learned in the Bay of Bengal, South Asia. Ecosystems and Livelihoods Group, Asia, IUCN, Colombo.
35. Von Essn, LM, SCA Ferse, M Glaser, and A Kunzmann. 2013. Attitudes and perceptions of villagers toward community-based mariculture in Minahasa, North Sulawesi, Indonesia. Ocean and Coastal Management 73: 101–112.
36. Yamamoto, T. 1995. Development of a community-based fishery management system in Japan. Marine Resource Economics 10: 21–34.
37. Yin, RK. 2009. Case study research, design and methods (4thed.). In Applied social research method series 5. Sage Publications, Los Angeles.

Chapter 8

The Material and Cultural Recovery of Camels and Camel Husbandry Among Sahrawi Refugees of Western Sahara

Gabriele Volpato and Patricia Howard

Department of Social Sciences, Wageningen University, Hollandseweg 1, Wageningen 6706, KN, the Netherlands.

ABSTRACT

For nearly 1,500 years, Sahrawi nomads of Western Sahara respected the camel; camels were essential to life in the desert environment, constituting both the main means of production and exchange and the keystone of Sahrawi cultural identity. The capacity to adapt to drought is crucial for the resilience of nomadic populations, which are particularly susceptible to its repeated occurrence. Knowledge of coping strategies is transmitted and embedded deeply within nomads' cultural institutions. In 1975, the Moroccan army occupied the Sahrawi's traditional nomadic territory, decimating camel herds and forcing most Sahrawi into refugee camps in Algeria where the Sahrawi became wholly dependent on foreign aid for their sustenance. However, with the signing of a ceasefire agreement in the early 1990s, the Sahrawi recovered part of their nomadic territory and the right to move within it, while at the same time, new flows of capital entered the camps. Refugees began to recover camel husbandry as a livelihood strategy and the camel re-emerged as a potent symbol as refugees and the Polisario Front (the Sahrawi's political representative) struggle to assert their new found national identity, regain access to all of their traditional territory and reaffirm their shared nomadic cultural heritage.

INTRODUCTION

For the past several decades, scholars and development organizations have increasingly focused on nomadic populations in part because, across the globe, their cultures and socio-ecological systems are threatened by the creation of political boundaries, forced sedentarization and resource depletion and in part because they have become aware of the complexity and efficacy of such systems in relation to resource management in harsh environmental conditions (Keenan[2003]; Gauthier-Pilters [1961]; Chatty [2005]). Research has also been conducted on the keystone role of the camel among these populations (Köhler-Rollefson [2003]), on their traditional ethnobiological and ethnoveterinary knowledge (Volpato et al. [2013a]; Mohamed and Hussein[1996]; Abbas et al. [2002]) and on the human-camel relations from an anthropological perspective (Catley [2006]).

Studies of herd loss and recovery among contemporary nomadic populations with drought and protracted conflict are central to pastoralist studies (Boneh [1984]; Khalif and Oba [2013]; Mezhoud and Oxby [2013]). Such studies are situated within an anthropological literature that understands sedentarization as a recurrent phenomenon within nomadic societies (Salzman[1980]; Adano and Witsenburg [2005]; Blench [2001]). The case of the Sahrawi can be contextualized precisely within this literature. Over the past 1,500 years, when their camel herds were largely depleted or severely weakened by drought or military assault, Sahrawi nomads could no longer appropriate desert resources and were forced to retreat to the periphery of their nomadic territories where they sought other means to survive outside of the desert, i.e. in oases, in coastal towns (where fishing was possible) or in agricultural settlements (Caro Baroja [1955]). Destitute nomads relocated to other areas either near wealthy camps or settlements (offering services to them, seeking assimilation and surviving on public slaughter and milk redistribution), or emigrating to the agricultural villages of the Oued Draa and Oued Nun or, less commonly, to the oases of central Mauritania. After herds were decimated by recurrent droughts, with the return of the rains, nomads began to repopulate the desert with their surviving herds or animals received in*mniha* (camel loans)[a]. Nomads who had lost all of their camels engaged in trade or other activities to acquire camels and initiate their real and metaphorical travel back towards the 'centre of the desert'.

Footnote

[a]Loans where those who borrow camels assume the costs of managing them and can use them for transport, milk production and wool.

In the 20th century, the strategies that the Sahrawi deploy to confront the Moroccan threat to their nomadic way of life and recover camels after dispossession while in exile in essence represent an extension or adaptation of their traditional mix of strategies for dealing with drought and military assault. In 1975, for the first time in their history, the Sahrawi were forced to confront an enemy with vastly superior military power and, in the process, lost all access to their traditional pastoral territories; most lost their mobility, herds and alternative means of livelihood. Most fled Morocco's military assaults, taking refuge in camps where they became utterly dependent upon foreign aid for the duration of the prolonged (16 years) guerrilla war that the Sahrawi waged against the Moroccan Army. The Sahrawi experienced greater social, cultural and political change living as refugees in a protracted situation. This time, the particular confluence of factors that permit refugees to proactively recover herds and reoccupy the desert, and the means by which they have adapted to these new conditions, signals profound changes in the economic, technical, social and cultural relations of camel husbandry in the Western Sahara and in the Sahrawi's social formation.

The study of pastoralists who have lost their herds and become refugees is directly relevant not only to the understanding of contemporary pastoralist societies that are confronted with herd loss and physical displacement, but as well to the study of refugees more generally and of their attempts to recover pre-conflict/pre-disaster livelihoods. Dispossessed pastoralists generally reconstitute herds only in the long term, which may be compared with the time required for refugees living in protracted situations to regain their subsistence strategies and cultural identity (Khalif and Oba [2013]; McCabe [1987]). In refugee camps, people's individual and collective agency - their capacity and actions to 'transcend the dictates of their immediate environment and... shape their life circumstances and the course of their lives' (Bandura [2006]. p. 165) - are severely constrained; they are deprived of access to most resources, confront limited opportunities (e.g. for mobility or education) and high levels of risk (e.g. looting, conflict, further displacement) (Jacobsen [2005]). When people become refugees, their traditional productive activities and the associated knowledge, cultural practices and artefacts may be lost forever. However, once in camps, as conditions permit, refugees struggle to reassert their agency, develop productive activities and revitalize the associated knowledge and practices. In the process, these are adapted to new environmental, social, cultural, political and economic contexts (Golooba-Mutebi and Tollman [2004]; Jacobsen [2002]; Horst [2006a]).

Because refugees lose everything with forced displacement into camps, recovery must occur from scratch. Studies about such autochthonous recovery processes (e.g. not purposively facilitated by aid agencies or governments) and of refugees' individual and collective agency can help to formulate policies that promote refugees' material and cultural well-being. The international community can support refugee's own efforts to regain access to the means of subsistence (e.g. livestock, land) by facilitating the economic means and necessary institutional conditions (e.g. mobility, security, legal entitlements) (Meyer [2006]). As the case study presented here shows, aid can be instrumental to the recovery process in that it represents a safety net for refugees who have no access to other resources, removes the burden of procuring basic provisions for survival while capitals are rebuilt and provides seed capital for refugees' own pursuits (Fiddian-Qasmiyeh

[2011]; Lentz et al. [2005]). As Harvey and Lind ([2005], p. 26) argue for Turkana pastoralists of Kenya, 'Food aid can provide the incentive for pastoralists to hold onto animals, and thereby contribute to the livelihood recovery process after an emergency.'

As refugees' worldviews, beliefs, values, cognition (including knowledge) and social relations are strongly rooted in their pre-exile worlds, their recovery strategies are informed by pre-exile modes of subsistence, social relations, cultural values and adaptive strategies. Especially among former pastoral and nomadic refugees, and within the constraints presented by the structure of camps and access to pastoral territories, refugees' agency is directed towards the recovery of livestock husbandry, including a return, as far as possible, to nomadism or transhumance. At the same time, however, refugees become integrated within trans-regional and transnational processes through political struggles, foreign aid, emigration and informal and formal markets for wage labour and other commodities. Pastoralism and nomadism are also being transformed through economic globalization (Chatty [2005]; Bollig et al. [2013]; Gertel and Le Heron [2011]), and refugee herd recovery is thus accompanied by change in the material and cultural use and management of herds related to these larger transformation processes. The individual and collective struggles to recover pre-exile modes of subsistence and livestock husbandry are motivated not only by the desire to improve the material conditions of life and gain greater autonomy; pre-exile modes of subsistence have great continuing cultural significance that refugees and their political representatives may re-appropriate and adapt to promote and strengthen identity that underlie claims to national sovereignty and territory.

In this paper, we address refugees' agency towards recovering camel husbandry, the conditions that permitted this process to occur and come to fruition, as well as the transformations in camel husbandry that are emerging, focusing on the case of Sahrawi refugees living in camps near Tindouf, in western Algeria[b]. Historically, camels were central to the production and reproduction of Sahrawi nomadic society, providing staple food, means of transport and warfare, means of exchange and the basis of power and prestige. With war and displacement, most of the Sahrawi who fled lost their camel herds, became refugees settled in camps and depended on international food aid for survival. Although several authors reported the recovery of livestock husbandry among Sahrawi refugees at the turn of the millennium (Bhatia [2001]; Caratini

[2007b]; Mundy[2007]; de Juan Canales [2010]; Herz [2013]), little attention has been given to the trajectory pastoralism took beginning with war and sedentarization in refugee camps or to Sahrawi refugees' agentic actions to recover camel husbandry. Indeed, the Sahrawi are seldom referred to in modern literature on pastoralism; for example, they are not mentioned in Homewood's ([2008]) recent book on the ecology of African pastoralist societies and are generally absent from edited books on pastoralism (Chatty [2005]; Bollig et al. [2013]). We argue that, at the beginning of the 1990s, a co-occurrence of factors (renewed access to territory, military demobilization and increased access to cash through pensions, remittances and aid) created conditions for the re-development of productive activities, trade and the emergence of an informal economy in the camps themselves. In continuity with their nomadic past, some refugees gained access to camels and resumed camel husbandry. These processes revitalized camel-associated traditional knowledge and the material and cultural bonds between refugees and their homeland (Volpato and Puri [2014]). The camel was re-appropriated as a symbol of a shared nomadic heritage and cultural identity, and camel products and associated knowledge and practices were deployed in the Polisario's and refugees' nationalistic discourse *vis-à-vis* Morocco as well as to reinforce national and ethnic unity. As discussed at the end of the paper, the fact that Morocco has also attempted to incorporate the camel into Moroccan national identity shows that there is a continuing struggle between the two parties to claim and monopolize Sahrawi nomadic heritage.

Footnote

[b]The data and information discussed in this paper are drawn from field-work carried out by the first author between 2006 and 2010 in the Sahrawi refugee camps and in the liberated territories of Western Sahara. Fieldwork included a survey (about camels owned, use of camels and camel products, extent to which the herd is used for market-oriented activities, camel management, distribution of camel products, personal history of camel husbandry; $n=74$) and semi-structured interviews ($n=36$) with refugee and nomadic camel owners, and a survey ($n=62$) with refugees' households about camel meat and milk procurement and consumption. Data were analysed using SPSS and Excel to derive descriptive statistics. The ethical guidelines followed were those adopted by the American Anthropological Association ([1998]) and the International Society of Ethnobiology ([2006])

After presenting the current context in the 'Background' section, in this paper, we: (1) describe how the Sahrawi became refugees and lost their herds; (2) discuss the historical, social, economic, cultural and ecological processes that facilitated the recovery of camel husbandry among Sahrawi refugees; (3) characterize contemporary camel husbandry and its importance within Sahrawi refugee camps; (4) discuss the changes that occurred in camel husbandry in the process and (5) analyse the contemporary role of camels in Sahrawi cultural and political identity.

BACKGROUND

The *Sahrawi*, literally 'people from the desert', are pastoralists who traditionally inhabited coastal areas of northwestern Africa including Western Sahara, northern Mauritania and part of southwestern Algeria. The Sahrawi were essentially nomadic, pasturing camels, goats and sheep in the low-lying plains of Western Sahara and relying for food on livestock products as well as dates, sugar, cereals and legumes that were bartered for livestock in markets on the periphery of their nomadic areas (Caro Baroja [1955]). In 1975, the Moroccan army occupied Western Sahara, forcing about 70,000 Sahrawi to flee and settle in camps established in neighbouring Algeria (San Martin [2010]). A 16-year war ensued between Morocco and the Sahrawi's armed political organization, the Polisario Front (1975 to 1991). Throughout this period, the Sahrawi were excluded from most of their former Western Sahara territory by means of a wall erected by the Moroccans (known as the *berm*) that cuts through Western Sahara in a north–south direction. Today, about 165,000 Sahrawi live in four refugee camps located on the Hamada desert plateau near Tindouf in Algeria (Figures 1 and 2).

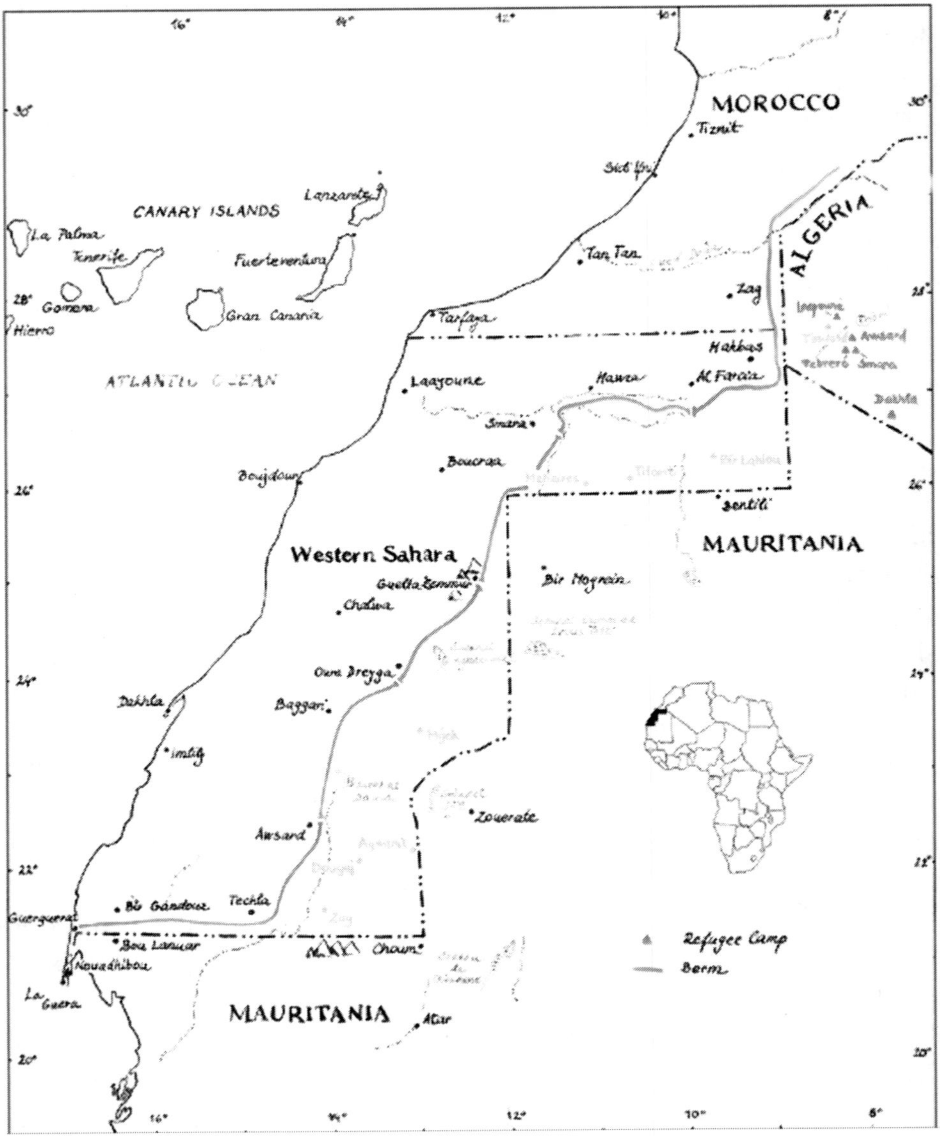

Figure 1: Map of Western Sahara with refugee camps and the liberated territories (P. Kourková).

Figure 2: Sahrawi refugee camps with livestock pens in the foreground (G. Volpato).

Refugees live in tents and mud brick huts and confront continuous problems with water, food and energy supplies. Food, shelter and other basic commodities are provided by the European Union, certain bilateral development programmes, UN agencies, the Algerian government and several solidarity groups (San Martin [2010]). Over the years, refugees have attempted to improve the quality of life in the camps, developing an informal economy where they market a variety of basic consumer products as well as food. They are involved in wage labour (e.g. as butchers, mechanics, construction workers, etc.) and trade between the camps and Mali, Mauritania, Algeria and Spain (Dedenis [2005]; Herz [2013]). Some have migrated abroad and send remittances home. Economic activity has greatly increased since a United Nations-sponsored ceasefire agreement was signed with Morocco in 1991, which resulted in the demobilization of Polisario troops who moved into the camps and began to re-engage in livestock husbandry, seasonal nomadism and trade.

With the ceasefire agreement, the Polisario Front also assumed political control over the Eastern, inland areas of the Western Sahara (Bhatia [2001]), the so-called liberated territories (approximately 20% of the Western Sahara)[c] while the remaining 'occupied territories' are under the Moroccan Government's administrative authority. Pastoral areas within the liberated territories are important to the refugees' struggle to maintain or recover traditional cultural and social practices, from livestock husbandry to medicinal plant use (Volpato et al. [2012]), as well as to generate income, e.g. through the sale of desert truffles (Volpato et al. [2013b]). Some 20 to 30,000 Sahrawi who remained nomadic throughout the war by moving with their herds to safe areas in neighbouring Mauritania and Mali also reoccupied these territories after 1991, herding their livestock and using the refugee camps and Zouérat (in Mauritania) as their main commercial hubs. Today, camel husbandry is practiced in the liberated territories as well as, to a lesser extent, in the areas surrounding the refugee camps. About 2,000 camels are raised in the camps, whereas some 40,000 are raised in the liberated territories (DNV-RASD [2007]). Furthermore, the Polisario own and tend some 27,000 camels in the liberated territories, where soldiers or paid labourers act as shepherds (Broglia and Volpato [2008]). In refugee camps, camel owners supplement natural forage with fodder purchased in Tindouf. In contrast, nomads (full-time and seasonal) practice extensive camel husbandry in the liberated territories, that is, husbandry that is highly dependent on pasture areas and mobility for herd production and reproduction.

Footnote

[c]See Volpato and Rossi ([2014]) for a photographic description of the liberated territories and of local nomads' life.

Whereas, throughout their history, tribal and camp leaders (re) distributed subsistence goods and means of production in camels and organized the defense of camps, camels and territory; today, the Polisario provides and distributes aid and defends and controls the refugee camps and the liberated territories. Although refugees are highly dependent on food aid, the Polisario produces a small part of the refugees' food supply. An important part of this production comes from Polisario livestock herds[d] which directly supply refugees with fresh camel milk and she-camels loaned to disadvantaged and poorer refugees, the elderly, former combatants and their families, and gift young or male camels for slaughter, e.g. for weddings. The Polisario also allocate milk-producing camels to workers in public places such as schools and hospitals. Polisario herds also supply livestock and their products for important rituals connected with social or political events. Livestock husbandry is one of the refugees' few endogenous activities developed and recovered without any consistent attention or funding from donors and development organizations. In the past decade, camels have also been the subject of international cooperation projects promoting health and nutrition through the supply of camel milk and meat (Mundubat [2007]). These projects are carried out and funded by non-governmental organizations (NGOs) under the Polisario's initiative and largely mirror the forms of support that the Polisario provides with its herds.

Footnote

[d]In 2005, the Polisario held about 40,000 sheep, 30,000 goats and 25,000 camels (DNV-RASD[2005]).

The recovery of camel husbandry is ongoing in the refugee camps of the Hamada and in the liberated territories where refugees travel and nomads live. The Hamada is a barren desert plateau known historically as the 'Devil's Garden' because rainfall is scarce and edaphic conditions are poor, which explains why it has very poor forage resources, few trees, no halophytic plants and few annuals. Across the liberated territories, the climate is also arid and continental, with summer daytime temperatures surpassing 50°C, while winter temperatures drop to zero at night. In the territories, rain generally occurs from the end of the summer through autumn. Average annual rainfall is 30 to 50 mm; however, rains are highly irregular both within and between years, and droughts are recurrent. But, in comparison with the Hamada, the liberated territories and especially the northern area, called Zemmur, are richer in biodiversity and forage plants. After the rains, they display a savannah-like environment dominated by *Acacia* trees, and flowering prairies may appear in flat gravel areas.

WAR, SAHRAWI NOMADS AND THEIR HERDS

Morocco's occupation of the Western Sahara and the ensuing war led to the decimation of camel herds and a drastic reduction in the number of Sahrawi that remained nomadic. In 1974, on the eve of the invasion, there were some 75,000 Sahrawi, only 40% to 45% of whom were nomads (Pazzanita [2006a], [b]; Censo-74 [1974])[e] tending about 50,000 camels, as well as numerous goats and sheep (Mercer [1976b]). Prolonged droughts in the 1960s and 1970s led many Sahrawi to sedentarize in the expanding urban areas in the periphery. The number of nomads and camels had been declining in Western Sahara since at least the 1950s, but this decline assumed almost epic proportions with occupation and war and might have led to the collapse, if not of this entire population, certainly of their way of life.

Footnote

[e]This census might have missed between 7 and 10,000 nomads (Gaudio [1993], p. 61).

Although figures are far from certain, it is estimated that about 55,000 Sahrawi (or about 70% of the population) fled to the refugee camps (Mercer [1976a]), about half of whom were nomads. The Moroccan air force bombarded temporary refugee camps in Western Saharan territory (at Oum Dreiga, Mahbes, Amgala and Guelta Zemmour) with napalm, white phosphorus and cluster bombs, causing Western Sahara towns to depopulate at the same time that Tindouf refugee camps swelled (Hultman [1977])[f]. Such panic was produced among most of the Sahrawi population that they tried to escape at any cost (San Martin [2005]), leaving their homes and belongings, livestock, jobs, in many cases, ceding their moveable property (herds, jeeps and trucks) to the Polisario forces that organized and carried out both the flight and the guerrilla war that ensued to recover their territory (Lippert [1992]; Brazier [1997]). The establishment of refugee camps behind the Algerian military outpost of Tindouf in Algerian territory meant that refugees were safe from bombings and that Polisario forces had a place to train former tribesmen and organize guerrilla war (Martin [1976]).

Footnote

[f]The Moroccan Army also poisoned wells to ensure that those people and herds that were not eliminated by the bombings would be unable to occupy the territory (Lippert [1987]).

Nomads, whose resilience and herds had already been weakened and reduced by droughts and colonialism (Capot-Rey [1962]; Mercer [1976b]), had to quickly decide whether to try to maintain their herds or liquidate them and flee to safety with other refugees. Apart from the bombings and well poisonings, they risked having their herds confiscated and being forced to sedentarize, having their mobility curtailed and dealing with possible lack of pasture within and across the borders of a country engulfed in war (Arkell [1991]; Gaudio [1993]). During the war, herds were bombed, abandoned, slaughtered and sold off by different agents (Pazzanita [2006a], [b]). The Moroccan offensive against camel herds was also meant to cut off the Polisario guerrilla's food supplies and logistical support. As a consequence, the large majority of Sahrawi nomads disbanded and joined the refugee camps soon after the invasion or within the next few years. Many attempted to sell their herds before fleeing in spite of the fact that prices had plummeted and cash buyers were difficult to find. At times, fleeing nomads abandoned herds in their customary nomadic territory in the (misplaced) hope that the animals could be recovered when stability had returned. Those who managed to reach refugee camps in Algeria with their herds often had to sell them (e.g. to local Algerian nomads) cheaply because they could not be fed and managed in the camps. Probably one of the most significant phenomena within this process was the Polisario's nationalization of camel herds. Some nomads turned their herds over to the Polisario rather than attempting to sell them, and the Polisario also appropriated some of the abandoned herds[g] and sought to manage 'national' camel herds within the nomadic territories that they could access or defend. These herds supported soldiers in the field (e.g. with meat and milk, since motor vehicles rather than camels were used for transport) and at times provided fresh meat for refugees.

Footnote

[g]There are no data available about the number of camels that Polisario obtained this way.

Still, several thousand nomads did not lose their herds or flee to the camps. A report from the early 1980s (Camacho [1987]) suggests that about 10,000 Sahrawi remained nomadic by moving to safe areas (e.g. West Algeria, Mauritania, northern Mali), making use of cross-border (or transnational) social relations based on tribal membership and solidarity with the Polisario. In the new system of social relations that emerged with the war, the Polisario became a supra-tribal power able to fight for the recovery of territory, represent the Sahrawi people as a whole internationally and provide refugees with the necessary means of consumption through political and humanitarian support. Nomads provided the Polisario militia with milk and meat from camels and small livestock and in turn were protected from Moroccan incursions. Those nomads that survived slowly moved back to the liberated territories after the ceasefire agreement of 1991.

Refugees suffered tremendously from the war: when fleeing through the desert, many lost children to starvation and spouses and parents succumbed to bombings, napalm attacks and extra-judicial executions (Lippert [1992]; Hernández [2013]). Old people, women and children suffered from severe malnutrition, lack of clean water, diseases and epidemic infections (Mercer [1976a]). In this context, it was impossible for tens of thousands of sedentarized refugees to rely on their own labour and resources for survival (Wilson [2012]). Providing the population with aid in the form of food, shelter, fuel and clothing was essential (Lippert [1992]). Those first years saw a collectivist war-economy develop based on international food aid managed and delivered to refugees by the Polisario and state (Sahrawi Arab Democratic Republic (SADR)) institutions[h] which established a national administration in exile in the refugee camps (San Martin [2005], [2010]). Refugees depended on the Polisario, which assumed and transformed many of the functions played by traditional nomadic Sahrawi camps and tribes (Farah [2008]). Irrespective of previous social differences (e.g. of wealth or caste), refugees all became equal in that they lost their livelihoods and property and became dependent on aid; livestock, the former basis of wealth, were nowhere to be seen. The few refugees who managed to bring livestock to the camps had to confine them in a single public pen (Wilson [2012]), effectively collectivizing their scarce resources. As Herz ([2013]), p. 12 argued, 'much of the nomadic culture was discarded, as life in the camps did not lend itself to the keeping of large camel herds.' Private trade was prohibited in the camps, and money was not used as currency. All of this began to change with the ceasefire agreement of 1991, which led to social relaxation in the camps, with greater freedom of movement of people and goods, the emergence of a private market sector and its acceptance by Polisario authorities and new opportunities for refugees to pursue individual and family livelihood strategies. This permitted Sahrawi refugees to exercise their agency in order to recover camel husbandry, as seen below.

Footnote

[h]The SADR is a partially recognized state that claims sovereignty over the territory of Western Sahara. It was established by the Polisario Front in 1976 to provide a new entity to fill the political void left by the departing Spanish colonizers. The SADR government controls about 20% of the Western Sahara territory.

REFUGEES' RECOVERY OF CAMEL HUSBANDRY

The recovery of camel husbandry among refugees was triggered by different but interacting economic, political, ecological, cultural and social drivers. The form this took was strongly informed by the past and also strongly influenced by the nature of these drivers and of the changing context. With the material and political change stemming from the ceasefire agreement of 1991 and, given the lack of alternative livelihood pathways in the desert environment, many refugees pursued camel husbandry as a path towards food security and livelihood diversification. The ceasefire agreement granted refugees access to part of their former nomadic territory (the liberated territories) and its grazing resources. Military operations ceased and Polisario soldiers were demobilized, which meant that many men with experience in camel husbandry returned to the camps and attempted to continue their activities. Sahrawi nomads also moved back to the liberated territories with camel herds that could be sold or loaned to refugees. Now firmly under the military control of the Polisario, safety and security in the camps and the liberated territories allowed refugees to invest and travel without threat to their persons and belongings and to renew mobility by means of jeeps, which allowed them to travel to markets (e.g. Tindouf) and to their former nomadic territories. Livestock and equipment purchases were facilitated by access to cash through, among other sources, Spanish civil pensions, donations and remittances. This also stimulated the development of productive activities and trade, leading to the growth of an informal economy in the camps, as well as widespread extra-camp trade and other economic and social interactions. Refugees emigrated (e.g. to Spain) and sent remittances or repatriated earnings while establishing transnational networks through diaspora. Camel and small livestock husbandry emerged as one of the very few means to sustainably exploit the resources of the Western Sahara, facilitated by the favourable rains of the past 10 to 15 years. Camel husbandry provides refugees with their traditional staple foods - fresh milk and, to a lesser extent, meat, in a context where, as in the past, agriculture is barely possible and hunting and gathering resources are very limited. Livestock husbandry traditionally provided resources for barter and now provides access to cash. Below, this process is addressed in more detail, starting with the means by which refugees gained access to camels in the camps.

Access to Capital and the Development of an Informal Economy in Refugee Camps

Soon after Polisario soldiers were demobilized and joined refugees in the camps, they began to engage in all sorts of informal private commercial and productive activity, boosting economic life. Up to that moment, mainly women undertook the limited economic activities in the camps as they attempted to supplement consumption; former soldiers slowly refocused household activities towards production for trade and commerce. A similar transformation occurred in relation to livestock: traditionally, women managed small livestock production for own consumption and they re-initiated this during the war, whereas camel husbandry and commercial livestock production can be traced in most instances to demobilized fighters (Mundy [2007]).

A large number of refugees gained access to cash through Spanish pensions, remittances, funds obtained through the programme *Vacaciones en Paz* (Holidays in Peace), formal and informal microcredit including loans from NGOs and other refugees, and trade (de Juan

Canales [2010]; Herz [2013]). In 1988, the Spanish Government authorized pension payments to about 5,000 Sahrawi who had worked for the colonial Government of the Spanish Sahara (e.g. as public workers and troops). These pensions (400 to 500€ per month) were paid retroactively as a lump sum, so many refugees received a substantial amount, which permitted them to invest in livestock and four-wheel drive vehicles as well as provide loans or credit to others, etc. Thus, cash began to circulate in the camps and incipient commercial activity arose with Tindouf and towns in Mauritania. Since the mid-1990s, a relatively steady number of Sahrawi refugees have emigrated to work in Spain (with smaller numbers going to other countries) to support their families in the camps, and emigration was often bankrolled by Spanish pensions. Remittances (e.g. from Spain, the Canary Islands, Algeria and Mauritania) also provided ready cash. A more limited flow of remittances and other forms of assistance enters refugee camps from Sahrawi living in that part of Western Sahara that is under Moroccan administration.

Another important source of cash comes from Spanish families who host between 7,000 to 10,000 Sahrawi children per summer in Spain through a programme known as *Vacaciones en Paz*(Crivello et al. [2005], [2006]; de Juan Canales [2009]). When returning to the camps, children bring gifts for themselves (e.g. clothes, toys) and their families (e.g. solar panels, appliances and money, on average 300 € per year) provided by the host families or local solidarity associations (Morando [2006]; Crivello et al. [2006]). Some host families collect money to send to the children's families in the camps, thus greatly contributing to refugees' income and to the emergence of a cash economy (Crivello et al. [2006]). It is estimated that this injects between 500,000 € and 1,000,000 € into the camps each year (Mundy [2007]).

Since the 1990s, refugees have also benefitted from small grants and microcredit programmes offered by NGOs to initiate small enterprises (Elizondo et al. [2008]), which has boosted productive activities and self-reliance among refugees (Cavaglieri [2005]). In some cases, microcredit has directly funded camel purchases to initiate milk production (López Belloso and Mendia Azkue[2009]). Besides formal microcredit, funds are provided informally by Sahrawi refugees through credit, loans and delayed payments in market transactions, particularly to relatives and acquaintances, increasing the number of refugees with access to cash. Cash empowered refugees through forms of redistribution, as is also found with other refugee populations (Koser and Al-Ali[2001]; Horst [2006b]). On a smaller scale, refugees with no access to cash may sell part of the aid that they receive to purchase other food (e.g. fresh camel meat and milk) or non-food products (e.g. coal) (Corbet [2008]), thus allowing them to participate in the camps' market economy.

The cash entering the camps permitted petty commerce, commodity production and a market economy to develop in the camps (Shelley [2010]; Herz [2013]) involving an ever-increasing number of refugees. Together, the potential to engage in long-term investment and commercial activities, and the proliferation of informal markets both within the camps and further afield, not only provided certain pre-conditions for the resurgence of camel husbandry but as well ensured that the social and technical relations entailed in newfound pastoralism would be substantially different from the nomadic past.

The Requirements of Camel Husbandry Recovery

Sahrawi refugees have attempted to recover livestock ever since they first settled in refugee camps. It has been shown in other contexts that one strategy used to restock is to diversify economically, as income gained in other sectors can be channeled into pastoralism (Horowitz and Little [1987]). Among the Sahrawi, the recovery began with small livestock (goats and sheep) and it was only after the mid-1990s that this evolved into camel husbandry. Today, a large majority of refugee households maintain small livestock in pens around the camps (Herz [2013]); compared with camels, small livestock are more economically accessible, have higher reproduction rates and fewer nutritional requirements, are more easily sold and have a more convenient size for own consumption; hence, pastoralists commonly begin with small livestock and only upscale to camel husbandry if and when conditions permit (Dahl [1979]; Mace and Houston [1989]). During the war, refugees generally lacked means to access camels (raids were banned and refugees had insufficient capital to purchase camels) and also lacked pastures in areas surrounding the camps. Accumulating the cash to purchase a camel (about 400 to 600 €) is not easy, and camel husbandry in the camps is even more onerous since further initial investments are required for pens, water tanks and fodder, while high risks of loss are entailed due to unfavourable husbandry conditions (Broglia and Volpato [2008]). The material requirements of camel husbandry included access to camels, to fodder and pastures, to associated social networks (e.g. for labour, herd management, product distribution, etc.) and also access to the requisite camel-related knowledge.

While the recovery of camel husbandry is still rather limited in terms of the numbers of refugees and camels involved, it is highly relevant both materially and culturally. While most refugee households own some goats and sheep, very few (1% to 5%) own camels, totaling only a few thousand camels per camp. For almost all such refugees, camel husbandry is a productive activity that complements food aid and meets social and cultural expectations related to both camels and their products. Camel husbandry continues to be constrained by refugees' lack of capital and by the unfavourable environmental conditions of the camps.

Diversity of Access to Camels

With the ceasefire agreement of 1991, access to the liberated territories and the demobilization of soldiers, on the one hand, and the increased inflow of cash into the camps, on the other, began to stimulate the recovery of camel husbandry, which was further sustained by the potential to market meat and milk in the camps. Camel purchases arose, in order of importance, in the liberated territories, Mauritania, Mali, Algeria and Niger, and were sometimes embedded in elaborate trade involving the Sahrawi diaspora. A number of cases demonstrate both the fact that the strategies refugees use to access camels may be quite complex, involving a 'chain' of activities and investments, as well as the diverse motives for accessing camels. Camels are gifted and inherited, as is the case with a 15-year-old boy who received a gift of two she-camels from his father. They are also purchased from savings - a 50-year-old refugee living in the Aaiun camp initiated a commercial camel herd by purchasing camels in Mauritania with cash obtained from his Spanish pension. Some are obtained through barter: one refugee emigrated to Spain, bought two cars and then exchanged them in the camps for eight camels, which are pastured in the liberated territories using a hired shepherd and which provide milk and other use values for his family when the rains come and they move to the liberated territories to live seasonally as nomads. One refugee migrated to Spain where he worked for a year in construction and, with the money he saved, bought a second-hand truck. He travelled with the truck via boat to Nouadhibou (Mauritania), and then overland to Azawad (in northern Mali), where he exchanged the truck for about 40 camels. He then travelled back to the camps and hired shepherds to bring the camels to him. Once there, the camels were sold to local butchers for a net profit. A woman who obtained remittances from her brother working in Spain bought a milk-producing camel to provide fresh milk for her widowed father. A couple with three children bought two she-camels with money that two of the children received from their hosts in Spain. Collective agency (requiring shared aims and coordination of interdependent plans of action; Bandura [2006]) may also be involved in camel purchases usually, but not always, among kin; in at least one case, four neighbouring (non-kin) households in the camp of Aaiun purchased a milk-producing camel, sharing the costs as well as the milk produced.

Other refugees obtained camels through non-monetary social relations (e.g. from nomadic or refugee kin or acquaintances with herds) as gifts (in which camel ownership is transferred) or, more often, loans (in which ownership is not transferred and camels are used for milk production). Reciprocity involving camels and camel products was very important to the maintenance of social cohesion among Sahrawi nomads and occurred mainly through kinship in accordance with tribal organization. Contemporary redistributive mechanisms serve functions that are similar to those strategies (e.g. *mniha* - the act of lending milk-producing camels and their calves to fellow tribesmen, an institution embedded in nomadic Sahrawi society) that redistributed camel surpluses through a wider population, permitting destitute nomads to reconstitute herds or survive drought or raids. Similar institutions are present in different pastoral populations throughout the world (Faye [2009]) and, in many cases, their demise has increased pastoralist vulnerability to droughts and other stresses (McCabe [1990]) or, conversely, they have become increasingly important to the survival and functioning of these communities (Ziker [2006]).

Once Sahrawi refugees have managed to renew access to camels, these traditional institutions have been revitalized and adapted: camels and camel milk began to be redistributed through gifts and loans, permitting access to a larger refugee population. One refugee camel owner, for example, gave three 2-year-old camels as gifts, one to a family of acquaintances, another to a young man who was helping him in construction work and a third to a cousin who gave the name of the camel owner to her newborn child. These camels may be then kept for milk production or, especially if young, slaughtered to provide meat for a wedding or baptism. However, the number of refugees receiving camels as gifts is rather limited. Loans (and delayed credit) are more widespread. *Mniha* plays an important role in the re-emergence of camel husbandry among refugees, where refugee and nomadic relatives from the liberated territories, Mauritania, and even from the Moroccan-controlled territories loan the camels, or camels may be loaned within the camps, as when a 'milk project' owner (see below) loaned a she-camel to the widow of one of his former workers.

Typology of Camel Husbandry in the Refugee Camps

We have addressed how refugees managed, through economic and social means, to access camels, arguing that this access was made possible by the inflow of cash into the camps and the re-activation of social networks. Renewed access to camels, however, did not give rise to the re-emergence of traditional extensive camel husbandry among refugees, but rather led to different forms of settled or partially settled husbandry. Refugee camel pastoralists are socially differentiated, which is reflected in herd size, social-economic relations and camel management practices. Broadly, three main forms of camel husbandry can be distinguished in the camps: (1) those who produce milk for own consumption (from two to a handful of she-camels); (2) small-scale commercial milk producers (from 10 to 30 camels), who sell milk after providing for own consumption and gifts; and (3) large-scale commercially oriented meat and milk producers (30 to 100 camels). In all cases, milk and meat are sold exclusively within the camps.

Refugees in the first category keep their camels under no-grazing conditions in a pen close to the household or on the camp's periphery (Figure 3). Camels are fed food scraps, hay and fodder and are watered from wells or potable water pipes. Refugees are strongly motivated to maintain camels for milk production (Figure 4) both because it is an essential staple and marker of cultural identity and because of beliefs that camel milk has health-giving and nutritional properties, especially for the ill and the elderly. These refugees consume or give away all milk produced. Milk is gifted to relatives or neighbours or given in exchange for labour (such as for camel milking). For example, soon after receiving his Spanish pension, one 60-year-old refugee and former nomad and his family moved to the periphery of the Smara camp, built a livestock pen and bought two she-camels with their calves. Although he lost his camels after a year, he re-invested and is now very proud of his four she-camels and three calves, which provide milk for his and one of his married son's family.

Figure 3: Camels in a pen (G. Volpato).

Figure 4: Camel feeding her calf in front of a refugee's house (G. Volpato).

Owners of medium-sized herds maintain most camels in free-grazing conditions in the liberated territories or in the Hamada of Tindouf, while some are kept in the camp to produce milk for own consumption, distribution and, at times, for sale. For example, one herder kept five milk-producing camels in the camp where besides providing for own consumption, a small fraction of the surplus milk produced was given away, while most was sold to cover the costs of fodder and water. This group includes many demobilized soldiers who maintain a 'traditional' approach to camel husbandry: they sell relatively small quantities of milk (e.g. 10 to 20 L per day), keeping some for themselves and for relatives and neighbours, while periodically distributing milk for free on a larger scale (some give all the milk produced on Fridays). The herd that is maintained in free-grazing conditions is usually loosely guarded if grazed around the camps or tended by nomadic relatives if grazed in the liberated territories. Such herds are used to support nomadic relatives, obtain cash from the sale of males and as a form of savings.

The third group consists of those who manage private enterprises of variable size (up to 100 camels) primarily devoted to sales of fresh camel milk (up to hundreds of litres per day) and/or meat in the camps. Their organization is more formal compared with smaller-scale producers and is based on the use of wage labour, with small quantities of milk destined for own consumption and gifted to the workforce. Milking is done by hand; milk is unpasteurized and is packaged in plastic bottles. It is taken to the camps by jeep and sold directly to refugees in local markets or indirectly via shops selling prepared food or groceries in the camps. Productive and non-productive she-camels are managed separately; milk-producing camels are kept on the periphery of the camps and fed mainly cut fodder, whereas the rest are kept under free-grazing conditions in the liberated territories or northern Mauritania. These enterprises are sometimes called 'milk production cooperatives' since some are the result of the joint investment of various refugees who then share the earnings. Milk projects have increased in number over the past decade due to increased camel milk demand. The profits obtained have generated substantial personal wealth for some. The main constraint to initiating a milk project is the lack of financial resources required to invest in a large milk herd; the funds used to date were obtained through remittances, Spanish pensions and/or earnings from other commercial activities.

Access to the Liberated Territories

The Sahrawi's traditional pastoral territory was almost inaccessible to pastoralists while engulfed in war from 1975 to 1991 but, since then, the 'liberated' areas have been key to the re-emergence of camel husbandry and camel-related practices among refugees as they provide grazing resources for refugees' and nomads' herds, as well as a place for nomads to live and refugees to move in seasonal or 'holiday' nomadism. Refugees' use of the liberated territories and northern Mauritania (where the Sahrawi have free informal access) for pasturing herds is informed by an economic rationale related to the emergent markets for livestock and livestock products in the refugee camps, as camel herds spend fallow or fattening periods in free-grazing conditions in these regions. This is especially the case for refugees with larger herds: in milk enterprises, fallow she-camels are grazed in the liberated territories as long as forage plants are available (after rains) to 'keep them healthy with good forage' and save the cost of feeding them in the camps on cut fodder. In meat enterprises, especially after rains, refugee herders use the liberated territories to fatten camels and small livestock before selling or slaughtering them in the camps.

Besides enjoying milk and meat in the camps, some refugees move seasonally with their livestock to the liberated territories where they consume these products. Some move to fatten livestock, purchasing small livestock and camels before heading for the *badiya*. One family bought about 300 goats and sheep and 40 camels which were sold upon return to the camps; in at least two cases, this system was embedded in the camel trade destined to supply meat to the camps - camels are bought in Mauritania or Mali, brought to the liberated territories where she-camels provide milk for seasonal nomadic refugees, males are fattened and, at the end of the period, sold to butchers in the camps.

Besides this economic rationale, seasonal nomadism is informed by many refugees' desire to enjoy a nomadic life in the *badiya* (Volpato and Rossi [2014]). Each autumn and/or winter, depending on the location of rains, thousands of refugees move to the *badiya* (especially its northern area) together with their livestock, tents, food stocks and various tools and other personal items, 'to graze camel herds and get away from the camps' (Mundy [2007], p. 295). Dedenis ([2005], pp. 86–87) and Caratini ([2007a], pp. 192–193) report that trips to the *badiya*may take the form of a 'holiday' (i.e. periods based on consumption rather than production), where refugees plan get-togethers with their nomadic relatives or celebrate important events (e.g. childbirths, weddings). Such seasonal nomadic strategies are encountered among other pastoral populations. In Ethiopia, for example, the refugee camps in neighbouring Somaliland were incorporated into seasonal nomadic movements so that nomads could rely on the camps' food networks during the dry season when pasture was difficult to find and milk was scarce, and refugees in turn enjoyed livestock products during the rainy season (Ryle [1992]). Boulay ([2004]) also described the phenomenon among former nomadic inhabitants of Nouakchott and other Mauritanian cities of a 'return to the *badiya*' after the rainy season for a milk and meat 'cure'.

Access to Knowledge

The knowledge a population holds about a mode of subsistence or productive activity is developed and transmitted across generations of engagement with a local environment and is commonly termed 'traditional', 'local' or 'indigenous' knowledge (Ellen et al. [2013]; Vandebroek et al.[2011]). As with other pastoral populations across the world, including the neighbouring Tuareg (Antoine-Moussiaux et al. [2007]), over the centuries, Sahrawi pastoralists have accumulated, shared and transmitted knowledge of camel husbandry in the desert environment of Western Sahara, including ethnobotanical knowledge about camel forage (Volpato and Puri [2014]), ethnoveterinary knowledge required to diagnose and treat camel diseases (Volpato et al. [2013a]) and ethnoecological knowledge about weather and rainfall patterns, soils, the location of wells and water points, distances and trajectories for transiting the desert and many other types of knowledge. With forced displacement, herd loss and sedentarization in refugee camps, knowledge transmission was disrupted. Younger people's disengagement with camels and the desert environment led to a lack of both the need and opportunities to learn. Decades of restrictions on access to nomadic territories meant there were no encounters with the learning environments in which camel and desert-related knowledge was transmitted, and knowledge acquisition and transmission became dormant (Volpato and Puri [2014]). A shift in values associated with formal education, emigration and exposure to mass media and to development schemes further alienated younger refugees from camel husbandry and the associated knowledge and cultural heritage, which some have even come to regard in negative terms.

With the resurgence of camel husbandry, which also involves young refugees who have no previous experience, many have struggled to revitalize the associated knowledge. The sources of this knowledge, principally older refugees and those who remained nomads, are involved in both vertical and horizontal knowledge transmission paths (Volpato and Puri [2014]). Recipients are not only those refugees who re-engage in camel husbandry but also political organizations and NGOs that support refugees. For the first two decades after exile, the SADR government and international organizations focused on guaranteeing the supply of basic items required for refugees' survival, and the conservation of traditional Sahrawi knowledge was not an issue of public concern. However, with the recovery of part of the Sahrawi's nomadic territory, the growing reconstitution of herds and flocks, and a string of years with favourable climatic conditions, camel-associated knowledge regained importance in the public sphere. Caratini ([2000], p. 445), for example, says that 'the Polisario Front have reconstituted a camel stock [...] and it realized that the loss of know-how in regard to livestock husbandry (based on the knowledge of the desert and of livestock needs) might represent a handicap both in the eventuality of a resumption of the war and given the perspective of peace.' In light of this, after the ceasefire agreement was signed, the Polisario involved its soldiers in a 'learning pastoral life' exercise involving nomadic shepherds (Caratini [2000]). The Polisario's revalorization of camel-related knowledge also entails the promotion of camel husbandry among younger generations and refugees as a symbol of cultural identity. For example, since 1992, the *Festival de la Cultura y de las Artes Populares* (Festival of Culture and Popular Arts) is held annually in Dakhla refugee camp, in which 'members of the older generations are actively engaged in teaching the youth a culture and traditional lifestyle that young people cannot find in school books' (SPS [2005]). Besides being crucial to camel husbandry, the recovery of camel-associated knowledge and values is functional to Polisario and Sahrawi refugees' assertions of political independence and rights to their territory as discussed below.

THE EMERGENCE OF PERI-URBAN, COMMODITIZED CAMEL HUSBANDRY

Technically and economically, the camel husbandry system developing in and around the refugee camps is quite similar to the peri-urban milk and meat camel husbandry systems that have developed around Saharan towns over the past few decades (Chaibou and Faye [2005]; Faye et al. [2003]). In fact, across arid and semi-arid areas of Africa, camel husbandry represents a viable productive pathway and has progressively become commoditized through marketing of camels and camel products in peri-urban husbandry systems. The transformation of pastoral nomads into petty commodity or capitalist producers, and of subsistence pastoral production into commodity production, has been occurring throughout Africa to various degrees and at various scales with increasing integration of camel milk and meat markets at regional and national scales (Herren[1992]; Nori et al. [2006]).

Recovery among Sahrawi refugees has thus entailed very substantial changes in environmental, technical and social relations of production and exchange compared with the nomadic past, and in several ways reflects refugees' dependence on, and engagement with, a transnational political economy. Their dependence on extra-territorial aid and income, their exposure to formal education and to formal and informal markets in the camps and abroad, their settlement in and reliance on the camps with limited access to their former territories and the fact that not all refugees are able to access the capital required to restock camel herds have all brought about such change. Below, we address three main aspects of this change, namely (1) the technological intensification of camel husbandry, (2) the commodification of camel products and (3) the establishment of new social relations around camel husbandry. We then address cultural expectations and resistance associated with these changes.

Changes in Camel Management and Technological Intensification

As the relationship with the nomadic territory disintegrated, camel husbandry in the camps came to rely far less on the biological resources of the desert and far more on imported capital and fodder. While nomadic camel husbandry depends on mobility and extensive free-grazing as main strategies to feed camel herds, refugees are sedentarized and lack grazing resources, so camel husbandry in the camps is mainly based on purchased fodder as well as flour, straw and barley grain. The economic viability of camel husbandry around the camps depends on the cost of feed and other inputs costs and access to the means to cover them. The need to purchase fodder places an economic burden on camel owners and often forces them to generate revenues from production to pay for feed; fodder is either purchased with the same sources of cash used to access camels or with revenue from milk sales.

The intensification of camel production among refugees does not occur through breeding or the use of tools or machines to substitute human labour (and thus reduce the costs of production). Rather, intensification, as understood here and as described for peri-urban camel systems in other parts of the world (Faye et al. [2003]), refers to the substitution of extensive land use (grazing on wild forage plants) with intensive input use - essentially, camels are confined in a smaller spaces and live from purchased fodder, trucked in water supplies and modern veterinary medicines. Different authors have noted a more general tendency in Saharan camel husbandry towards reduced mobility and grazing intake and increased consumption of fodder and supplementary feeds (Michel et al. [1997]; Faye et al. [2003]). Throughout the last century or so, the use of supplementary livestock feeds has increased worldwide in part as a result of the increasing availability of cheap agro-industrial by-products and mechanized transport systems capable of delivering them in bulk (Blench [2001]).

In most cases, fodder is purchased in Tindouf. Travel and trade, including for fodder purchase, are facilitated by the diffusion of vehicle transport that, since the 1990s, greatly facilitated refugees' mobility (Dedenis [2005]). After the war, jeeps and trucks have superseded camels as a means of transport. Camels now provide transport only on a limited scale and only among nomads, but few nomadic households regularly move to new grazing areas with their belongings packed on camels. Although, as Bulliet ([1975]) described so brilliantly, for 2,000 years, the camel managed to displace the wheel in North Africa and the Middle East, in the 20th century, the wheel, now coupled with the internal combustion engine, took its revenge, and the use of camels for transport went into steep decline. The widespread use of jeeps and trucks in desert environments has had profound impacts on camel pastoralists' lifestyle (Chatty [1980], [1986]). Owning a 'land rover' is a principle ambition among Sahrawi refugees and one of the first commodities they acquire when they have sufficient cash is a motor vehicle. These facilitate trade and inter-camp transit, serving as taxis and transporting goods to and from Tindouf, the liberated territories and Mauritania. People ride camels in the camps only during SADR political or ritual celebrations when mounted camel parades are performed to demonstrate the Sahrawi's nomadic origins (Figure 5). The camel has become a symbol of Sahrawi cultural identity in public and political rituals and demonstrations as, for example, in 1997, when James Baker, the former UN mediator of the Polisario-Morocco conflict, visited the camps and again in 2006, in Tifariti, for the 30th anniversary of the SADR's foundation.

Figure 5: Camel riding during a political celebration in Rabouni refugee camp (G. Volpato).

The diffusion of motor vehicles has also changed camel management. Most successful refugee camel owners own a jeep or truck. Especially market-oriented camel herders use other complementary technologies such as tanker trucks, GPS and satellite phones. Trucks make it easier to supplement camels' diets with harvested wild plants, hay and fodder. The historical constraints related to water provision for camels are addressed by developing new wells, mechanization (pumps) and tanker trucks that bring water to the herds, which obviates the need to move herds to water. There is generally increased reliance on owned or rented tanker trucks both among refugees and nomadic herders, particularly for larger, market-oriented herds. Herds can exploit grazing resources in the liberated territories that would otherwise be too distant from water points, whereas in the camps, it permits large herds to be maintained on the periphery where water is deposited in large tanks.

Camel Milk Commodification

Although historically, the Sahrawi traded surplus camels, an unprecedented process of commodification of camels and camel products is occurring among refugees. Camels are no longer bartered for food and other essentials but are sold for profit; meat is sold in butcher shops and milk, which it was formerly taboo to sell, is subject to widespread trade and high consumer demand. As trade gave rise to an informal economy in the camps, camel milk has become at least partly commoditized, as have other natural products that were previously not subject to monetary exchange, such as plant medicines and desert truffles (Volpato et al. [2013b]).

Studies investigating Sahrawi refugees' dietary intake confirm that camel milk is widely consumed (Henjum et al. [2010]; Grewal [2011]). Refugees source camel milk in three ways: (1) own production, which is very limited due to the small number of refugees who own or maintain camels; (2) social mechanisms of redistribution in which refugee camel owners, the Polisario Front and certain international NGOs maintain milk-producing camels in and around the camps and provide refugees with fresh milk through family, social and/or political networks; and (3) purchase in camp shops and directly from refugee producers, which accounts for the majority of the milk consumed in the camps. About half of the refugee camel owners surveyed by the first author ($n=44$) sell camel milk. Nomads ($n=30$), however, do not sell camel milk but rather, as in the past, redistribute it through kinship and other networks, to neighbouring households that have no milk-producing camels and to labourers that manage herds as well as to their families.

The commodification of camel milk in Sahrawi refugee camps reflects trends reported in urban markets across Saharan Africa (Faye et al. [2003]), from the Western Sahara under Moroccan administration (Michel et al. [1997]) and Mauritania (Abeiderrahmane [2005]) to Kenya (Noor et al.[2013]) and Somalia (Farah et al. [2007]); the same trends are found, for example, in Saudi Arabia (Faye et al. [2014]). Commodification occurs through two possible paths: through the penetration of capitalist relations of production, and/or as a necessary adaptation to husbandry in sedentary conditions. The second particularly applies to Sahrawi refugees; camels had to be fed purchased fodder. Sahrawi refugees were forced to incorporate fixed and variable capital to substitute resources otherwise obtained for free in their traditional territories. In order to obtain the capital needed to maintain camels in the camps, refugees began to sell at least part of their milk. In a few but far-reaching cases, this led some refugees to develop large herds and seek profits. At the same time, these changes have been met with resistance on the part of some refugees with a nomadic background who challenge changes that have occurred such as the commodification of camel milk, which has fundamentally altered social relations and challenged traditional values and beliefs. Indeed, the Sahrawi have maintained a more traditional approach in nomadic conditions (e.g. in the liberated territories), where the redistribution of camel milk still constitutes the basis of the nomadic community, giving credibility to the idea that it is need, rather than greed, that has led to milk commodification.

Social Relations around Camel Husbandry

Over the past 50 years, Sahrawi social relations and political organization have undergone no less than a revolution. Major social and political change commenced with colonialism and war and culminated with exile and the creation of a Sahrawi nation-state, the SADR (see footnote h). Historically, the Sahrawi were organized politically in tribes, where several inhabited the traditional territories (e.g. Reguibat, Izarguien, Oulad Delim, Oulad Tidrarin, etc.; see Briggs

[1960]; Pazzanita [2006a], [b]) in a segmentary system that further subdivided each tribe into branches based on extended families and other kinship ties (Hart [1998]; Caro Baroja [1955]). Tribal relations determined access to pasture where each tribe controlled a customary nomadic area. Decision-making institutions and networks of mutual support functioned along tribal lines and especially within branches (Caratini [1989]). The tribes of Western Sahara continuously interacted through trade, war, raids, agreements of mutual access to each other's customary areas (e.g. with drought), alliances, tribute payments, etc. Although this political organization was disrupted with colonialism, it was in the mid-1970s that it was completely overturned by events: with exile and the establishment of the camps, the Polisario banned tribes and reorganized the Sahrawi under a newly declared nation-state in which each refugee became a citizen of the SADR and pledged allegiance to the Polisario, which controls and grants refugees access to the liberated territories and manages food aid (Wilson [2012]; Caratini [2003]). While kinship continues to have considerable importance among contemporary Sahrawi, tribal affiliation gave way to Polisario allegiance, and the Polisario Front de facto assumed the functions previously assumed by tribes and extended this to incorporate other functions of the modern nation-state, e.g. international political representation. Among Sahrawi refugees, social relations have been established and/or reformulated around camel husbandry on the basis of kinship (e.g. between refugees and nomads) and on Polisario governance (e.g. for access to the liberated territories).

The process transforming tribal authority to a nation-state is exemplified by the changes that occurred in camel brands. Branding refers to cauterizing unique symbols (marks) on an animals' body to communicate information about ownership, the owner's social status and grazing rights (Landais [2001]; Humphrey [1974]). Brands are also essential elements of identity in pastoral societies, as their collective character serves as an assertion of collective rights (e.g. of tribes) to specific resources (e.g. water, pastures) (Landais [2001]; Bernus [1996]). Among Sahrawi nomads, the process of branding camels was known as *mhar* (*nar* in Arabic) or *elama*; camels were usually branded on the neck or the thigh, or on the head or other visible part of the body (Caro Baroja [1955]; Boyer [1962]). Each tribe had its own brand, while larger tribes (e.g. the Reguibat in the 19th century) had different brands for major segmentary branches. Monteil ([1952]) recorded more than 25 brands used by Sahrawi tribes or their bands, about half of which incorporated letters or combinations of letters from the Arabic alphabet (e.g. the letters *kaf* and*qaf* for the Reguibat), the rest consisting of stylizations or abstract symbols.

The origins of brands were usually embedded in myths that tied into a single narrative the tribe's founder, the camel and the brand (Caro Baroja [1955], pp. 89–90). Camel brands conveyed the power of a tribe not only in terms of the number of camels held but also the location of camel herds in the territory - an indicator of territorial control. The brand, then, was a message conveyed to other pastoral groups about the related priority of access (or not) to the territory where those camels grazed. Thus, for pastoralists, the brand can be considered as a functional and symbolic equivalent of national flags for claiming territorial occupation and defining which human groups are granted access.

In the social revolution from a tribal to a state organization led by the Polisario, brands were also revolutionized. Today, all camels belonging to Polisario are marked on the neck with the letters 'FP' (Frente Polisario) followed by a number relating to a military region (from one to seven). For Polisario herds, this brand has completely displaced tribal marks, including a noticeable shift in the alphabet from Arabic to Latin, and a shift in the symbolic function of these herds' identity, from one based on tribal divisions to one based on Sahrawi unity. The Polisario adopted the branding conventions of the colonial authorities - Spanish and French colonial powers branded their herds with numbers rather than symbols. Polisario brands thus represent a rejection of traditional tribal symbols, reinforcing the ideological unity of the Sahrawi people beyond tribes and draw upon colonial symbols of state ownership. However, in continuity with the tribal past, in the liberated territories, the presence of camels marked with 'FP' indicates control over those territories and its resources and hence to whom people owe allegiance when using those resources.

With the recovery of camel husbandry and renewed access to a shared nomadic territory, refugees began to interact more intensively with nomads, often on the basis of kinship ties. Some nomads and their herds became integrated with the camel-based economy emerging in the refugee camps by functioning as reservoirs of camels (e.g. through loans) and camel-related knowledge and as a source of camel milk for refugees engaging in seasonal nomadism. Nomads in turn use the refugee camps as a commercial hub where they sell livestock, procure other items, access aid through refugee relatives and exchange food with refugees. As Wilson ([2012]) also noted, both refugees and nomads use the liberated territories to graze herds, interacting in a vast network of solidarity and reciprocity that is strengthened by kinship ties linking nomadic and refugee households. The common interest in camel husbandry has largely defined their relationships, served to strengthen social bonds that were severed with the dissolution of tribes and contributed to the development of a system of apparent mutual benefit.

The Sahrawi case is not the first in which refugee camps are shown to have a positive impact on regional economies by functioning as commercial hubs for local nomads who sell their surplus animals in the camps. Jansen ([2011]) reports the same for the Kakuma camps of Kenya. There, however, nomads and refugees are from different and rather antagonistic ethnic groups; the former are Turkana and the latter mainly Dinka, so that the relations between them are strictly based on trade. Among the Sahrawi, there is a deep interpenetration between refugees and nomads, so that patterns of exchange involve extended families and reciprocity, in addition to market exchange.

In the past, labour relations around camel husbandry were based on extended families and on structural differences within and between tribes. While poorer nomads relied on own and male kin labour, more powerful nomads and tribes (i.e. with hundreds of camels) exploited the labour of slaves and dispossessed nomads (e.g. members of a defeated tribe) (Caro Baroja [1955]; Hart[1962]). Today, nomads and refugees with small non-commercial herds still rely principally on family labour. However, in market-oriented enterprises, labour relations around camel husbandry have been largely commoditized, representing a break from the types of structural social differences that prevailed in the past. Commercially oriented herders use hired labour for camel management (e.g. feeding and watering) in and around the camps and contract shepherds to tend herds in the liberated territories. In the camps, labour is mainly provided by young refugees who have little formal education. In the liberated territories, refugee youth, Sahrawi nomads (local or from Mauritania) or, in some cases, sub-Saharan migrants are contracted. Shepherds are paid in cash rather than in kind whereas, in the past, they were paid in livestock, e.g. a camel in exchange for a year's labour.

Cultural Expectations and Resistance around Camel Husbandry

Refugees do not share the same attitudes towards the recovery of camel husbandry and renewed access to camel products, nor to the changes in camel husbandry and camel management. The refugee literature identifies two contrasting processes of cultural change in situations of exile (Couldrey and Morris [1999]; Volpato and Waldstein [2014]). The first is a tendency towards loss or abandonment of traditional culture, which is maintained only among elderly and specific, often marginal, groups. The second is a process of maintenance or recovery of traditional culture in order to promote the maintenance or recovery of productive activities and cultural identity. Among Sahrawi refugees, those who have been at least partly acculturated in the West and many other younger refugees who have never experienced nomadic life often express less concern about cultural continuity and regard camels and camel husbandry as 'backward' unless they are used as a means to make money. Older and more culturally conservative refugees stress the importance of livestock and livestock products, especially camel milk, and demand that their children meet their expectations by consuming camel milk and meat and travelling to the liberated territories.

Incentives to recover camel husbandry include the desire, especially of older refugees, to consume fresh camel milk and spend at least part of their lives in the *badiya*. Dedenis ([2005]) writes that sons continue to be motivated by the social obligation to provide fresh livestock products for their parents and to reconstitute herds; feelings of good health, pride, dignity and freedom are associated with a 'return to the *badiya*'. For the elderly, travelling through space to the *badiya* represents travel in time back to their youths. The *badiya* and its key elements (green deserts after rains, livestock, camel milk, tents) are emblematic of the culture they represent - of nomadic Sahrawi - and of associated values such as freedom and autonomy[i].

Footnote

[i]An association between 'freedom' and the '*badiya*' is often found in nomadic Sahrawi poetry, which was largely based on the description and celebration of *badiya* resources and landscapes. For example, the most renowned living Sahrawi poet, Mohamed Mustafa Salem (known as Badi) said, 'In the *badiya*a man is free to do everything he pleases, with his life, with his livestock, with his water. He is owner of himself. Who does not live in the *badiya* cannot be owner of himself, in any other place...Every minute of life one feels free, very free. Nothing is better in the *badiya* than when a friend comes, a person you esteem and to whom you can offer everything you can imagine' (cited in Gimeno Martín and Pozuelo[2010], pp. 11–12) [own translation from Spanish].

Forms of cultural resistance arose among refugees around technological change and intensification, where some are against the use of tanker trucks to water camels, which 'corrupts nomadic life'. Because traditional camel pastoralists have internalized camels' needs and the constraints of camel husbandry into their systems of knowledge and beliefs about the natural world, in their eyes, such change goes against the laws of nature. Some believe that, although tanker truck water comes from wells, it is not as good for camel's health as water that is directly extracted from wells and that camels watered directly with well water have greater strength and stamina[j]. Similarly, some traditional herders are 'against fodder' and are proud of the fact that they rely exclusively on free-grazing resources, believing that purchased fodder is bad for camels' health and lowers their resistance. The dialectic between 'innovation' and 'tradition' in camel management has, in some cases, assumed the form of an intergenerational struggle in which older men tend to favour traditional strategies. These forms of cultural resistance are probably an integral part of situations where profound change in the mode of subsistence and social relations are underway and have also been shown to occur among Sahrawi refugees in relation to concepts of illness, health and healing (Volpato and Waldstein [2014]).

Footnote

[j]According to Elmi ([1989]), Ceeldheer pastoralists of Somalia do not like to water their camels from wells with a motorized pump for two reasons: the camels do not like the water because it smells of diesel, and water stored in metal tanks is cooler than that from the bottom of the well, and thus, camels do not like it and will not drink it to capacity.

THE CAMEL AND SAHRAWI IDENTITY

All the people like us are we, and everyone else is they (Rudyard Kipling, We and They, 1926).

Kohler-Rollefson ([1996], p. 283), in her work on the origin, use and dispersal of the one-humped camel in Asia, writes, 'The function of animals in human societies is usually not restricted to utilitarian purposes; they often fulfil important and culturally distinct symbolic and ritual roles...and quite frequently develop into hallmarks of identity.' Historically, among the Sahrawi, the camel was not only the material pillar of society but as well the cornerstone of Sahrawi tribal and nomadic identity. Several Sahrawi tribes' origin myths were based on camels (Caro Baroja [1955]; Mercer[1976a]); camels were central to Sahrawi beliefs, sayings and poems (Pinto Cebrián [1997]); camel ownership defined prestige and social rank; and nomads identified with and felt affection for their camels and their local camel breeds. With recovery, much of the cultural significance of the camel as a symbol of ethnic identity has been revitalized, which is reinforced by the Polisario's efforts to create and sustain a national identity based on ethnicity, cultural heritage and territorial occupation.

Identity refers to feelings of belonging to a specific group based on a distinctive shared language, culture, social organization and ideology (Hurskainen [1990]; Snow [2001]). Identities and their symbols are continuously constructed and re-constructed by redefining the differences between the self and the other, the cultural borders in which these differences are represented, and abandoning or adopting different key elements of identity (Drzewiecka [2002]; Nagel [1994]). As Corbet ([2008]) noted, contemporary Sahrawi are struggling in multiple ways to reconstruct their identity. The camel is a central, figurative and symbolic element that represents the Sahrawi and differentiates them from other (non-camel pastoralist) populations. The re-establishment of a relation with the *badiya* based on pastoral and nomadic practices reaffirms Sahrawi refugees' cultural identity based on their nomadic origins.

When the Polisario overturned tribal social structure, it eliminated the basis of social differences in favour of national unity, social equality and Sahrawi collective identity and citizenship and, at the same time, the cultural and political significance of camels was undermined. This revolution was necessarily cultural as well as social; tribalism in all its forms, as well as the material basis of this tribalism, were antithetical to the Polisario's project of building the basis for nationalist claims to the territory against other nationalist claims (e.g. Morocco). The Polisario thus not only eliminated (prohibiting reference to) tribal institutions and affiliations but as well discouraged the transmission of tribal myths of origin, which affirmed the distinct genealogical identities of the Sahrawi tribes as well as the symbolic significance of the camel as the material and cultural basis of these tribes' existence. As the Polisario nationalized camel herds, the camel was purged of any significance as the basis of tribal or personal power in favour of a cultural identity based on the unity of all ethnic Sahrawi under a 'modern' nation-state. However, the camel was too potent a symbol both of Sahrawi ethnic identity and of Sahrawi claims to their traditional nomadic territory to discard wholly or permanently. The camel was later to be recovered as a symbol of Sahrawi unity *vis-à-vis*other populations and so regained a prominent place in Sahrawi cultural identity not only among refugees and nomads engaged in camel husbandry but as well in official Polisario discourse and practices. For the Polisario, the rebirth of the camel's symbolic significance was functional to claims of independence. Camel husbandry reinforces the Polisario's claims to nation-statehood by legitimizing control over the former territory pertaining to the Sahrawi nomadic tribes both through occupation (people and animals) and use and by providing for the national welfare and reducing dependence on foreign aid by means of state food production and redistribution. The camel and camel husbandry also serve to legitimize claims to national identity by representing the basis of a shared ethnic, cultural and political identity, as well as symbolizing the differences between the Sahrawi and other populations, particularly the Moroccans.

The contrasting political ideologies and claims of the Polisario, on the one hand, and of the Morocco Government, on the other hand, constitute a crucial axis of the dispute over control of Western Saharan territory. Under the Polisario, Sahrawi identity has been reinforced on the basis of differences between the Sahrawi and the Moroccans, whose cultural practices are explicitly and implicitly contrasted (Chamberlain [2005]). The SADR's Ministry of Information produces educational and informational material for refugees that stress this contrast in terms of language, food, dress and women's roles (Allan [2010]) and, as a result, 'many aspects of everyday Saharawi existence become cultural markers' (Chamberlain [2005], p. 27). This is also the case with ethnobotanical knowledge associated with the desert environment and camel husbandry (Volpato and Puri [2014]), which contrasts with Morocco's agricultural traditions. The dispute over Western Sahara territory thus aims to 'hegemonise a collective social imaginary about what it is to be Saharawi, who the Sahrawis are and who the 'others' are that delineate the frontier of 'our Sahara' (San Martin [2005], p. 587). Camels and camel husbandry are part of this dispute and are central to the public representation of Sahrawi political identity. Camels appear in refugees' and Polisario's discourses as the basis of a 'stolen' nomadic life in the *badiya*, where 'pastures are good' and 'camels gain health', in contraposition to neighbouring areas (e.g. Morocco to the north, central Mauritania to the south) where 'camel diseases are widespread' and 'the climate is too wet for camels'. Below, we present three examples of this re-appropriation of the camel as a symbol of cultural and national identity in relation to the figure of the *deyar* (a nomad who searches for stray camels), to the skills needed to slaughter camels and to the use of camel milk in political contexts.

In Sahrawi nomadic culture, the *deyar* is a very knowledgeable and experienced nomad camel tracker who locates camels that have strayed in the desert and returns them to their herds. *Deyars* read tracks in the sand and infer a great amount of information from their shape, depth, distribution, direction, etc. *Deyars* are highly respected for their knowledge, skills and intimate relation with the desert environment. With the collapse of nomadism during the war and the changes that ensued in camel management (e.g. increased reliance on motor vehicles and binoculars), the number of *deyars* drastically declined and their skills are being lost. But if their physical presence is dwindling, their cultural significance is not: on the contrary, *deyars* have a prominent place in present-day Sahrawi literature and poetry (see for example Ebnu [2008]; Boicha et al. [2003]; Bahia [2007]; Moure [2004]), where they often represent the *badiya* itself and the lost relationship with and knowledge of Western Sahara territory. At the same time, the*deyar's* search for stray camels has come to symbolically represent the Sahrawi's search for the camels lost with the war and, in a wider sense, of the search for their lost nomadic life, independence, and freedom in the territory of Western Sahara, and hence an assertion of the claim to the territory.

Among the cultural practices that the Sahrawi contrast explicitly with the Moroccans are camel slaughter skills, which reflect the intimate relation between camels and the Sahrawi as well as the alleged Moroccan's ignorance of camels. The contrast is stressed by a young Sahrawi author who describes a camel slaughterhouse on the outskirts of El Aaiún city in Moroccan-controlled Western Sahara, where camels can be heard crying out in suffering, which he contrasts with Sahrawi practices known as *nosh* (Autores Varios [2009], pp. 89–90)[k]. One of the *nosh* prescriptions dictates that, when slaughtering a camel, suffering must be minimal, which is possible only through the skills and experience of the slaughterer, which are transmitted between herders across generations as part of Sahrawi cultural heritage. So, for example, when a young Sahrawi attended an encounter with King Hassan II in 1979 organized by the Moroccan authorities, they sacrificed a camel: it 'was horrible, that animal was crying a lot... those people don't have any idea of how a camel must be killed' (A.A. [1979]). Indeed, if the main arteries between the neck and the thorax are severed with a single cut, death occurs instantaneously, whereas if the cut is not expertly executed, death is delayed and the animal suffers (Farah and Fischer [2004]). In pre-colonial Sahrawi society, usually, the eldest or most respected man slaughtered camels, which was forbidden to lower social groups (e.g. slaves). Thus, a contraposition that was internal to the social system (slaves-camel owners) has been conceptually reorganized as an external contraposition (Moroccans-Sahrawi), which is functional to Polisario political struggles and claims to Sahrawi unity.

Footnte

[k]*Nosh* literally refers to tethering camels by grabbing them by the tail.

Since at least 1999, the struggle for Sahrawi nationhood has extended into that part of Western Sahara under Moroccan control where the Sahrawi experience social and economic discrimination and participate in nationalistic demonstrations and riots (San Martin [2005]). In this context, camels are an issue; a group of former nomads from El Aaiún is currently campaigning for compensation for the camel herds that Moroccan troops killed during the war, which requires the Moroccan authorities' recognition that troops were ordered to butcher Sahrawi herds, a fact that Moroccan authorities deny (Shelley [2004]). Such acts are contrasted with the Sahrawi's affection for the animal and the social and the cultural importance attached to camels and camel products. Sahrawi who struggle for independence in the Moroccan-controlled Western Sahara adopt camel milk as a symbol of their ethnic and national identity while reflecting the difference with Moroccans' sedentary lifestyles and identities. Reports show that, in April 2006, when Moroccan authorities released Sahrawi activists from prison and again in Agadir, in 2007, when Sahrawi students were released for having publically demanded independence for Western Sahara, their Sahrawi supporters greeted them with cups of camel milk and dates (SPS [2006], [2007]), just as Sahrawi nomads returning from long desert treks were received in their camps in pre-colonial times.

Morocco is also promoting camel husbandry in the part of Western Sahara under its control while attempting to appropriate the camel as a symbol at a sub-national (i.e. ethnic, tribal) level, thus proposing that Sahrawi nomadic heritage is part of Moroccan national heritage (Martin-Márquez[2006]; Julien [2004]). Polisario calls for Sahrawi unity *against* tribal identity, while Morocco tries to promote and enhance tribal identity and downplay or deny the existence of Sahrawi national unity (Corbet [2008]). Moroccan authorities are attempting to undermine the Polisario's monopoly over the politicization of Sahrawi nomadic culture, exalting the camel and its characteristics. This has occurred in the context of a parallel recovery of camel husbandry in Moroccan-controlled Western Sahara which has been on-going since the 1990s, especially with the emergence of commercial enterprises producing and selling camel milk and meat (Pazzanita [2006a], [b]; Michel et al. [1997]). The Moroccan Government and local authorities increasingly fund camel husbandry, supporting it through direct aid (e.g. subsidies for camel purchases), cooperation programmes that distribute free veterinary medicines and vaccines and implement applied research (e.g. epidemiological studies) (El Abrak [2000]; Essemlali [2008]; Michel et al. [1997]). Today, the Moroccan Government and media portray camel husbandry as an integral part of the Moroccan economy and promote the camel milk and meat consumption (El-Katab [2008]). Among other events promoting camel meat and milk in Western Sahara, the Moroccan authorities organized an International 'Salons of the Dromedary' in 2008 and 2009 in what they call the 'Southern Provinces' or 'Moroccan Sahara' to celebrate 'the central symbol of the culture and life of Saharan inhabitants', as the francophone Moroccan newspaper *Le Matin du Sahara et du Maghreb*(Essemlali [2008]) reported: a clear example of the attempt to reposition Sahrawi nomadic heritage as part of Moroccan heritage. Thus, a struggle to ideologically appropriate Sahrawi nomadic cultural heritage continues between Moroccan authorities and the Polisario, with the former aiming to incorporate it as a specific ethnic/tribal heritage *within* Moroccan national culture and the latter aiming to recover nomadic cultural heritage as the foundation of Sahrawi independence and against Moroccan claims[I].

Footnote

[I]See Wallerstein ([1987]) for a discussion of the political manipulation of concepts of identity as 'national', 'ethnic' or 'racial' where 'peoplehood' is 'in no sense a primordial stable social reality, but a complex clay-like historical product of the capitalist world economy through which the antagonistic forces struggle with each other' (p. 85), which seems entirely appropriate in the context of Western Sahara.

CONCLUSIONS

This study has addressed the material and cultural importance of camels and camel husbandry among Sahrawi refugees, as well as their efforts to recover their territory and productive activities and promote their shared nomadic heritage and cultural identity. The case illustrates both how refugees struggle to recover their livelihoods in contexts of forced displacement and sedentarization in refugee camps and how this struggle is informed by pre-exile modes of subsistence and cultural values. The camel was central to Sahrawi nomadic society both materially and culturally. With war and forced displacement, the Sahrawi lost most of their camels - it took decades for refugees to recover camel husbandry by regaining access to camels, grazing territory and camel-associated knowledge and to establish reciprocal relations around camel husbandry in the liberated territories with remaining nomads. Although this recovery is sustained by traditional values and practices related, for example, to camel milk consumption and redistribution, fundamental changes have also occurred: the camel lost its traditional functions as a means of transport, warfare and non-monetary exchange; management has intensified with the use of fossil fuels and motor vehicles and increasing reliance on purchased fodder and other inputs rather than on desert grazing resources; and camel products as well as access to camels and labour have been largely commoditized. The camel has also been appropriated as a symbol of Sahrawi cultural identity both by refugees and in Polisario discourse and practice, which asserts the Sahrawi's nomadic cultural heritage as a key element of a national identity legitimating claims to nationhood and traditional territories for the Sahrawi people.

The findings of this study are significant for understanding both refugees' individual and collective agency towards cultural and economic recovery and dispossessed pastoralists' struggle to rebuild herds and livestock-based forms of livelihood. We have expanded our knowledge of Sahrawi refugees' agency and their re-engagement with pastoralism and the liberated territories and also provided a framework for more generally conceptualizing some of the processes by which refugees living in camps engage in productive activities, how these are informed by their culture, access to means of production and financial capital and how they are embedded within larger-scale socio-economic change processes including commodification and migration. The findings also contribute to the understanding of human-nature relations in contexts of forced displacement and of the means by which refugees re-engage with traditional productive activities, knowledge and environmental relations, in the process adapting (and reinventing) the past to new conditions and values arising from their experience as refugees. Knowledge of these processes is important not only to understand refugees' and displaced people's struggles but also to comprehend broader cultural and social processes taking place in refugee camps (and, within this, the role of food aid) and to formulate ways in which the international community can address refugees' problems within these processes. NGOs working in refugee camps can use studies about the recovery of refugees' livelihoods to identify and implement means to assist this recovery.

Although the recovery of camel husbandry is rather limited in terms of the numbers of refugees and camels involved, it is likely to continue into the future as refugees continue to break the chains of food dependency, as long as it can be sustained by rains, and as long as access to camels and to the liberated territories remain assured. The strategy that refugees pursue leads towards greater integration of refugee camps with wider economic and social networks, but this depends on the continued existence of the camps and of international aid. As long as this trend continues, there will be an increasing number of camels and greater integration between refugees and nomads (often within extended families), creating a single productive network between the refugee camps and the liberated territories. If the referendum for Sahrawi self-determination is eventually held[m] and refugees move back to Western Sahara, some refugees might return to nomadism, merging with nomads on both side of the berm and using Western Sahara towns as commercial hubs. In this scenario, the Polisario Front might redistribute its camels to refugees, in a sort of concluding cycle in which the herds that the Polisario acquired from disbanding nomads at the beginning of the war and used to support refugees in the decades of exile, once again become the basis for a return to desert nomadic pastoralism.

Footnote

[m]The ceasefire agreement of 1991 included a referendum for self-determination originally scheduled for 1992, in which the Sahrawi would have the option to decide between independence and integration with Morocco. However, this referendum was never held and is still in the UN's to-do list (Pazzanita [2006a], [b]).

AUTHORS' CONTRIBUTIONS

GV designed the methods approach, carried out field work, analysed the data, composed the literature review and drafted the manuscript. PH participated in the study design, in the literature review and in drafting the manuscript, as well as editing it. Both authors read and approved the final manuscript.

ACKNOWLEDGEMENTS

We are very grateful to many Sahrawi for the patience, kindness and hospitality that they showed during fieldwork. We thank the Italian NGO Africa 70 - especially Alessandro Broglia, Sara Di Lello and Davide Rossi - and the Veterinary Service of the SADR (Sahrawi Arab Democratic Republic) and its director Saleh Mohamed Lamin for logistical support and cooperation. Our gratitude also extends to the Polisario personnel who accompanied Gabriele Volpato in his travels across the liberated territories of Western Sahara. Funds were granted to Gabriele Volpato by Ceres Research School of Wageningen University, The Netherlands, as part of his PhD research on the subsistence practices and ethnobiological knowledge of Sahrawi refugees.

REFERENCES

1. 1979. La extraña aventura de Chefid Mohamed. El Pais, Madrid.
2. Abbas, B, AA Al-Qarawi, and A Al-Hawas. 2002. The ethnoveterinary knowledge and practice of traditional healers in Qassim Region, Saudi Arabia. Journal of Arid Environments 50: 367–379.
3. Abeiderrahmane, N. 2005. Modern Dairy Products from Traditional Camel Herding: An Experience in Mauritania. In Desertification Combat and Food Safety, Series I: Life and Behavioural Sciences, vol. 362, ed. Faye B, Esenov P, IOS, Amsterdam.
4. Adano, WR, and K Witsenburg. 2005. Once Nomads Settle: Assessing Process, Motives and Welfare Changes on Marsabit Mountain. In As Pastoralists Settle: Social, Health, and Economic Consequence in Marsabit District Kenya, ed. Fratkin E, Roth EA, 105–136. Kluwer Academic/Plenum, New York and London.
5. Allan, J. 2010. Imagining Saharawi women: the question of gender in POLISARIO discourse. The Journal of North African Studies 15 (2): 189–202.
6. 1998. Code of Ethics of the American Anthropological Association.
7. Antoine-Moussiaux, N, B Faye, and GF Vias. 2007. Tuareg ethnoveterinary treatments of camel diseases in Agadez area (Niger). Tropical Animal Health and Production 39: 83–89.

8. Arkell, T. 1991. The decline of pastoral nomadism in the Western Sahara. Geography 76 (331): 162–166.
9. Bahia, MA. 2007. Versos Refugiados. Poesía Saharaui Contemporanea. Universidad de Alcala, Alcala de Henares.
10. Bandura, A. 2006. Toward a psychology of human agency. Perspectives on Psychological Science 1 (2): 164–180.
11. Bernus, E. 1996. Marques de propriété touarègues et pierres tombales (plaines au sud-ouest de l'Aïr). Sahara 8: 7–18.
12. Bhatia, M. 2001. The Western Sahara under Polisario control. Review of African Political Economy 28: 291–298.
13. Blench, R. 2001. 'You can't go Home again': Pastoralism in the new Millennium. Overseas Development Institute, London.
14. Boicha, L, L Lehsan, S Ebnu, S Abdalahi, AS Iselmu, and C Mahmud. 2003. Bubisher. Poesía Saharaui Contemporánea. Puentepalo, Las Palmas de Gran Canaria.
15. Bollig, M, M Schnegg, and H Wotzka. 2013. Pastoralism in Africa: Past, Present and Future. Berghahn, Oxford.
16. Boneh, D. 1984. Returning to pastoralism: three cases from the Negev Bedouin. Nomadic Peoples 15: 41–50.
17. Boulay, S. 2004. Quand un Objet Change de Statut: Trajectoire de la Tente Dans la Société Maure (Mauritanie). Ethnographiques. org 6, .
18. Boyer, JR. 1962. Contribution à l'étude de l'élevage Camelin au Sahara Occidental. Le Regueibat et son Chameau. Ecole Nationale Veterinaire D'Alfort, Paris.
19. [http://www.newint.org/features/1997/12/05/light/] Brazier, C. 1997. Light in the Darkness. : New Internationalist magazine (297).
20. Briggs, LC. 1960. Tribes of the Sahara. Harvard University Press, Cambridge.
21. Broglia, A, and G Volpato. 2008. Pastoralism and displacement: Strategies and limitations in livestock raising by Sahrawi refugees after thirty years of exile. Journal of Agriculture and Environment for International Development 102 (1/2): 105–122.
22. Bulliet, R. 1975. The Camel and the Wheel. Harvard University Press, Cambridge.

23. Camacho, A. 1987. Pastores en la guerra. El Pais, Madrid.
24. Capot-Rey, R. 1962. The Present State of Nomadism in the Sahara. The Problems of the Arid Zone. UNESCO, Paris.
25. Caratini, S. 1989. Les Rgaybat (1610–1934). 2: Territoire et Société. L'Harmattan, Paris.
26. Caratini, S. 2000. Système de parenté Sahraoui. L'impact de la révolution L'Homme 154–155: 431–456.
27. Caratini, S. 2003. La Republique des Sables: Anthropologie d'une Revolution. Editions Harmattan, Paris.
28. Caratini, S. 2007. La prison du temps. Les mutations sociales à l'oeuvre dans les camps de réfugiés sahraouis Deuxième partie: l'impasse Afrique contemporaine 2 (222): 181–197.
29. Caratini, S. 2007. La prison du temps. Les mutations sociales à l'oeuvre dans les camps de réfugiés sahraouis. Première partie: la voie de la révolution. Afrique contemporaine 1 (221): 153–172.
30. Caro Baroja, J. 1955. Estudios Saharianos. Consejo Superior de Investigaciones Cientificas, Madrid.
31. Catley, A. 2006. Use of participatory epidemiology to compare the clinical veterinary knowledge of pastoralists and veterinarians in East Africa. Tropical Animal Health and Production 38: 171–184.
32. Cavaglieri, S. 2005. Livelihood Strategies and Food Security in Refugee Camps. University of Roma Tre, Rome.
33. 1974. Censo, ed. Censo y Estadistica Servicio de Registro de Poblacion. Servicio de registro de poblacion, censo y estadistica, El-Ayoun.
34. Chaibou, M, and B Faye. 2005. Fonctionnement des élevages camelins de la zone périurbaine d'Agadez au Niger: enquête typologique. Revue d'Élevage et de Médecine Vétérinaire des Pays Tropicaux 58 (4): 273–283.
35. Chamberlain, R. 2005. Stories of a Nation: Historical Narratives and Visions of the Future in Sahrawi Refugee Camps. Refugee Studies Centre, University of Oxford, Oxford.
36. Chatty, D. 1980. The Pastoral Family and the Truck. In When Nomads Settle: Processes of Sedentarization as Adaptation and Response, ed. Salzman P, 80–94. Praeger, New York.

37. Chatty, D. 1986. From Camel to Truck: The Bedouin in the Modern World. Vantage, New York.
38. Chatty, D. 2005. Nomadic societies in the Middle East and North Africa: Entering the 21st century. In Handbook of oriental studies. Section 1 The Near and Middle East. Brill, Leiden.
39. Corbet, A. 2008. Nés Dans les Camps. Changements Identitaires de la Nouvelle Génération de Réfugiés Sahraouis et Transformation des Camps. École des Hautes Études en Sciences Sociales, Paris.
40. Couldrey, M, and T Morris. 1999. Culture in Exile. Forced Migration Review. The Refugee Studies Centre in association with the Norwegian Refugee Council/Global IDP Project, Oxford.
41. Crivello, G, E Fiddian, and D Chatty. 2005. The transnationalisation of care: Sahrawi refugee children in a Spanish host program. Refugee Studies Centre, Oxford.
42. Crivello, G, E Fiddian, and D Chatty. 2006. Holidays in peace: Sahrawi children visit Spain. FMR 25: 59–60.
43. Dahl, G. 1979. Suffering Grass: Subsistence and Society of Waso Borana. Stockholm Studies in Social Anthropology. University of Stockholm, Stockholm.
44. de Juan Canales, D. 2009. Imaginación y reacción en los Campamentos de refugiados saharauis: construcción de la cotidianeidad e identidad frente al otro extranjer (nasrani) en el contexto de la diáspora. Perifèria - Revista de recerca i formació en antropologia 10: 1–24.
45. de Juan Canales, D. 2010. Emergencia civil y economía del refugio en los campamentos de refugiados saharauis de Argelia. Gazeta de Antropologia 26 (1): 11.
46. Dedenis, J. 2005. La Mobilité Régionale de la Société Sahraouie des Camps de Réfugiés. Une Ressource Pour un Mieux-Vivre?. Université de Poitiers, Poitiers.
47. DNV-RASD. 2005. Census 2005. República Árabe Democrática Sahrawi, Rabuni, Polisario refugee camps: Dirección Nacional de Veterinaria, Ministerio de Salud Pública Sahrawi, RASD.
48. DNV-RASD. 2007. Census 2007. República Árabe Democrática Sahrawi, Rabuni, Polisario refugee camps: Dirección Nacional de Veterinaria, Ministerio de Salud Pública Sahrawi, RASD.

49. Drzewiecka, JA. 2002. Reinventing and contesting identities in constitutive discourses: between Diaspora and its others. Communication Quarterly 50 (1): 1–23.
50. Ebnu, A. 2008. Nómadas en el Exilio. Asociación Cultural Almenara, Marbella.
51. El Abrak, A. 2000. Encadrement Sanitaire du Cheptel Camelin au Maroc. In Maladies Parasitaires et Infectieuses du Dromadaire, ed. Dakkak A, 9–14. Actes Editions, Rabat.
52. Elizondo, L, CM Mehdi, and M Sanz. 2008. Microcréditos en el Sáhara. Manual Para la Solicitud de Crédito en los Campamentos Saharauis. Hegoa, Bilbao.
53. El-Katab, A. 2008. La viande du dromadaire, un produt sahraoui jusque-lá inexploité. Libération, Rabat.
54. Ellen, R, SJ Lycett, and SE Johns. 2013. Understanding Cultural Transmission in Anthropology: A Critical Synthesis. Berghahn Books, Oxford.
55. Elmi, AA. 1989. Camel Husbandry and Management by Ceeldheer Pastoralists in Central Somalia. Botany and Range Science Department, Faculty of Agriculture, Somali National University, Mogadishu.
56. Essemlali, M. 2008. Provinces du Sud: Le dromadaire symbole central dans la culture sahraouie. Le Matin, Rabat.
57. Farah, R. 2008. Refugee camps in the Palestinian and Sahrawi national liberation movements: A comparative perspective. Journal of Palestinian Studies 38 (2): 76–93.
58. Farah, Z, and A Fischer. 2004. Milk and Meat from the Camel: Handbook of Products and Processing. vdf Hochschulverlag AG, Zurich.
59. Farah, Z, M Mollet, M Younan, and R Dahir. 2007. Camel dairy in Somalia: limiting factors and development potential. Livestock Science 110: 187–191.
60. Faye, B. 2009. Pauvreté et Solidarité Chez les Peuples Pastoraux. In L'élevage, Richesse des Pauvres, ed. Duteurtre GB, Faye B, 77–87. QUAE, Versailles.
61. Faye, B, M Bengoumi, and A Barkat. 2003. Le Développement des Systémes Camélins Laitiers Périurbains en Afrique. In Lait de Chamelle Pour l'Afrique, 115–125. Niamey: Atelier sur la Filière

Laitiére Caméline en Afrique Niamey, 5–8 November, 2003. FAO Production et Santé Animales, Niamey.

62. Faye, B, H Madani, and AH El-Rouili. 2014. Camel milk value chain in Northern Saudi Arabia. Emir J Food Agric 26 (4): 359–365.
63. Fiddian-Qasmiyeh, E. 2011. The pragmatics of performance: Putting 'faith' in aid in the Sahrawi refugee camps. Journal of Refugee Studies 24 (3): 533–547.
64. Gaudio, A. 1993. Les Populations du Sahara Occidental. Editions Karthala. Paris:
65. Gauthier-Pilters, H. 1961. Observations sur l'ecologie du dromedaire dans le Sahara Nord-Occidental. Mammalia 25 (2): 195–280.
66. Gertel, J, and R Le Heron. 2011. Economic Spaces of Pastoral Production and Commodity Systems: Markets and Livelihoods. Ashgate, Farnham.
67. Gimeno Martín, JC, and LM Pozuelo. 2010. Memorias orales en el Sáhara Occidental: la poesía en hasanía. 7th Congress of African Studies. Lisboa 21: 1–27.
68. Golooba-Mutebi, F, and SM Tollman. 2004. Survival to livelihood strategies for Mozambican refugees in South Africa. FMR 20: 28–29.
69. Grewal, NK. 2011. Iodine Status and Thyroid Function Among Lactating Women in Saharawi Refugee Camps, Algeria. Akershus University College, Akershus.
70. Hart, DM. 1962. The social structure of the Rgībāt Bedouins of the Western Sahara. Middle East Journal 16 (4): 515–527.
71. Hart, DM. 1998. The Rgaybat: Camel nomads of the Western Sahara. The Journal of North African Studies 3 (4): 28–54.
72. Harvey, P, and J Lind. 2005. Dependency and Humanitarian Relief: A Critical Analysis. Humanitarian Policy Group, Overseas Development Institute, London.
73. Henjum, S, I Barikmo, AK Gjerlaug, AM Lehabib, A Oshaug, TA Strand, and LE Torheim. 2010. Endemic goitre and excessive iodine in urine and drinking water among Saharawi refugee children. Public Health Nutrition 13 (9): 1472–1477.

74. [http://www.elmundo.es/elmundo/2013/10/20/internacional/1382244113.html] Hernández, R. 2013. La Semilla de la Verdad. : Sáhara Documental. . Accessed 20 Oct 2013.
75. Herren, UJ. 1992. Cash from camel milk: the impact of commercial camel milk sales on garre and gaaljacel camel pastoralism in southern Somalia. Nomadic Peoples 30: 97–113.
76. Herz, M. 2013. From Camp to City: Refugee Camps of the Western Sahara. ETH Studio Basel, Lars Müller, Zürich.
77. Homewood, K. 2008. Ecology of African pastoralist societies. James Currey Ltd., Oxford.
78. Horowitz, MM, and P Little. 1987. African Pastoralism and Poverty: Some Implications for Drought and Famine. In Drought and Hunger in Africa: Denying Famine a Future, ed. Glantz M, Cambridge University Press, Cambridge.
79. Horst, C. 2006. Refugee livelihoods: Continuity and transformations. Refugee Survey Quarterly 25 (2): 6–22.
80. Horst, C. 2006. Transnational Nomads. How Somalis Cope With Refugee Life in the Dadaab Camps of Kenya. Studies in Forced Migration. Berghahn Books, New York.
81. Hultman, T. 1977. The struggle for Western Sahara. Issue: A Journal of Opinion 7 (1): 27–32.
82. Humphrey, C. 1974. Horse brands of the Mongolians: a system of signs in a nomadic culture. American Anthropologist 1: 471–488.
83. Hurskainen, A. 1990. Levels of identity and national integrity: the viewpoints of the pastoral Maasai and Parakuyo. Nomadic Peoples 25–27: 79–92.
84. 2006. ISE Code of Ethics (with 2008 additions).
85. Jacobsen, K. 2002. Livelihoods in conflict: the pursuit of livelihoods by refugees and the impact on the human security of host communities. International Migration 40 (5): 95–123.
86. Jacobsen, K. 2005. The Economic Life of Refugees. Kumarian, Bloomfield.
87. Jansen, BJ. 2011. The Accidental City: Violence, Economy and Humanitarianism in Kakuma Refugee Camp, Kenya. Wageningen University, Wageningen.

88. Julien, Z. 2004. L'identité Sahraouie en Questions. In L'Ouest Saharien, ed. Abjean A, Julien Z, 131–231. L'Harmattan, Paris.
89. Keenan, J. 2003. The last nomads. Nomadism amongst the Tuareg of Ahaggar (Algerian Sahara). Journal of North African Studies 8 (3–4): 163–192.
90. Khalif, ZK, and G Oba. 2013. Gaafa dhaabaa - the period of stop': Narrating impacts of shifta insurgency on pastoral economy in northern Kenya, c. 1963 to 2007. Pastoralism: Research, Policy and Practice 3: 1–14.
91. Kohler-Rollefson, I. 1996. The one-Humped Camel in Asia: Origin, Utilization and Mechanisms of Dispersal. In The Origins and Spread of Agriculture and Pastoralism in Eurasia, ed. Harris DR, 282–292. UCL, London.
92. Köhler-Rollefson, I. 2003. A treatise on life with camels. German Research 1: 15–19.
93. Koser, K, and N Al-Ali. 2001. New Approaches to Migration? Transnational Communities and the Transformation of Home. Routledge, London.
94. Landais, E. 2001. The marking of livestock in traditional pastoral societies. Revue scientifique et technique de l'Office international des épizooties 20 (2): 445–462.
95. Lentz, E, CB Barrett, and J Hoddinott. 2005. Food aid and dependency: implications for emergency food security assessments. IFPRI Discussion Paper 12(2).
96. Lippert, A. 1987. The human costs of War in Western Sahara. Africa Today 34 (3): 47–59.
97. Lippert, A. 1992. Sahrawi women in the liberation struggle of the Sahrawi people. Signs 17 (3): 636–651.
98. López Belloso, M, and I Mendia Azkue. 2009. Local human development in contexts of permanent crisis: Women's experiences in the Western Sahara. Journal of Disaster Risk Studies 2 (3): 159–176.
99. Mace, RH, and AI Houston. 1989. Pastoralist strategies for survival in unpredictable environments: a model of herd composition which maximises household viability. Agricultural Systems 30.
100. Martin, P. 1976. Tribesmen swell Polisario ranks. The Times, London.

101. Martin-Márquez, S. 2006. Brothers and others: fraternal rhetoric and the negotiation of Spanish and Saharawi identity. Journal of Spanish Cultural Studies 7 (3): 241–258.

102. McCabe, J. 1987. Drought and recovery: livestock dynamics among the Ngisonyoka Turkana of Kenya. Human Ecology 15 (4): 371–390.

103. McCabe, J. 1990. Success and failure: the breakdown of traditional drought coping institutions among the pastoral Turkana of Kenya. Journal of Asian and African Studies 25 (3–4): 146–160.

104. Mercer, J. 1976. The cycle of invasion and unification in the Western Sahara. African Affairs 75 (301): 498–510.

105. Mercer, J. 1976. Spanish Sahara. G. Allen, London.

106. Meyer, S. 2006. The 'refugee aid and development' approach in Uganda: Empowerment and self-reliance of refugees in practice. New Issues in Refugee Research. Research Paper 131. :Geneva.

107. Mezhoud, S, and C Oxby. 2013. Disentangling 'Forced displacement' from Pastoral Mobility: Recovery and Reconstruction in the Sahel and in South Sudan. In Modern Pastoralism and Conservation: Old Problems, new Challenges, ed. T Sternberg and D Chatty, 210. The White Horse, Cambridge.

108. Michel, JF, M Bengoumi, P Bonnet, K Hidane, K Zro, and B Faye. 1997. Typologie des systèmes de production camelins dans la province de Laâyoune, Maroc. Revue d'Élevage et de Médecine Vétérinaire des Pays Tropicaux 50 (4): 313–323.

109. Mohamed, HA, and AN Hussein. 1996. Pastoralists: their knowledge untapped. Journal of Camel Practice and Research 3: 143–145.

110. Monteil, V. 1952. Essai sur le Chameau au Sahrara Occidental. Etudes Mauritaniennes. Centre IFAN - Mauritanie, Saint-Louis du Senegal.

111. Morando, M. 2006. La Diaspora Saharawi in Italia. Universitá degli Studi di Bologna, Bologna.

112. Moure, G. 2004. La Zancada del Deyar. El Cobre, Barcelona.

113. Mundubat,. 2007. Acta de la II Mesa de Coordinación y Concertación. In Mesa de Concertación de Soberanía Alimentaria del Sahara. Madrid :.

114. Mundy, JA. 2007. Performing the nation, pre-figuring the state: the Western Saharan refugees, thirty years later. Journal of Modern African Studies 45 (2): 275–297.

115. Nagel, J. 1994. Constructing ethnicity: creating and recreating ethnic identity and culture. Social Problems 41 (1): 152–176.

116. Noor, IM, AY Guliye, M Tariq, and BO Bebe. 2013. Assessment of camel and camel milk marketing practices in an emerging peri-urban production system in Isiolo County, Kenya. Pastoralism: Research, Policy and Practice 3(28).

117. Nori, M, MB Kenyanjui, MA Yusuf, and FH Mohammed. 2006. Milking drylands: the marketing of camel milk in North-east Somalia. Nomadic Peoples 10 (1): 9–28.

118. Pazzanita, AG. 2006. Western Sahara: anatomy of a stalemate. International Journal of African Historical Studies 39 (1): 151–154.

119. Pazzanita, AG. 2006. Historical Dictionary of Western Sahara. Historical Dictionaries of Africa. The Scarecrow, Oxford.

120. Pinto Cebrián, F. 1997. Proverbios Saharauis. Miraguano Ediciones, Madrid.

121. Ryle, J. 1992. Notes on the repatration of Somali refugees in Ethiopia. Disasters 16 (2): 160–168.

122. Salzman, PC. 1980. When Nomads Settle: Processes of Sedentarization as Adaptation and Response. Praeger, New York.

123. San Martin, P. 2005. Nationalism, identity and citizenship in the Western Sahara. The Journal of North African Studies 10 (3): 565–592.

124. San Martin, P. 2010. Western Sahara: The Refugee Nation. University of Wales Press, Cardiff.

125. Shelley, T. 2004. Endgame in Western Sahara: What Future for Africa's Last Colony?. Zed Books, London.

126. [http://www.redpepper.org.uk/sons-of-the-clouds/] Shelley, T. 2010. Sons of the Clouds. Accessed 2 Sept 2011.

127. Snow, DA. 2001. Collective Identity and Expressive Forms. In International Encyclopedia of the Social and Behavioral Sciences, ed. Smelser NJ, Baltes PB, 2212–2219. Elsevier Science, Oxford.

128. SPS. 2005. Segunda Jornada del Festival: Inauguración del Mukkayem Tradicional. Rabouni, Tindouf, Algeria. Accessed 29 Apr 2006.

129. 2006. Recibimiento Triunfal a Tamek en Assa. Assa: Sahara Press Service, Rabouni, Tindouf, Algeria.

130. 2007. Acogida Triunfal a los Estudiantes Saharauis en la Universidad Ibn Zohr de Agadir (Marruecos). Sahara Press Service, Agadir.

131. Vandebroek, I, V Reyes-García, UP Albuquerque, R Bussmann, and A Pieroni. 2011. Local knowledge: Who cares? Journal of Ethnobiology and Ethnomedicine 7(35).

132. Varios, A. 2009. La Fuente de Saguia. Relatos de a Generación de la Amistad Saharaui. Imprenta Provincial de Zaragoza, Zaragoza.

133. Volpato, G, and R Puri. 2014. Dormancy and Revitalization: The Fate of Ethnobotanical Knowledge of Camel Forage among Sahrawi Nomads and Refugees of Western Sahara. Ethnobotany Research & Applications, Accepted.

134. [http://www.smashwords.com/books/view/433218], Volpato, G, and D Rossi. 2014. Badiya: Los Nómadas Saharauis del Sáhara Occidental, ed. Smashwords.

135. Volpato, G, and A Waldstein. 2014. Eghindi among Sahrawi refugees of Western Sahara. Med Anthropol: Cross-Cultural Studies in Health and Illness 33 (2): 160–177.

136. Volpato, G, P Kourková, and V Zelený. 2012. Healing war wounds and perfuming exile: the use of vegetal, animal, and mineral products for perfumes, cosmetics, and skin healing among Sahrawi refugees of Western Sahara. Journal of Ethnobiology and Ethnomedicine 8 (49): 1–19.

137. Volpato, G, A Di Nardo, D Rossi, SM Lamin Saleh, and A Broglia. 2013a. 'Everybody knows', but the rest of the world: the case of a caterpillar-borne reproductive loss syndrome in dromedary camels observed by Sahrawi pastoralists of Western Sahara. Journal of Ethnobiology and Ethnomedicine 9(5).

138. Volpato, G, D Rossi, and D Dentoni. 2013. A reward for patience and suffering - Ethnomycology and commodification of desert

truffles among Sahrawi refugees and nomads of Western Sahara. Economic Botany 67 (2): 147–160.

139. Wallerstein, I. 1987. The construction of peoplehood: racism, nationalism, ethnicity. Sociological Forum 2 (2): 373–388.

140. Wilson, A. 2012. Households and the production of public and private domains: revolutionary changes in Western Sahara's liberation movement. Paideuma 58: 19–43.

141. Ziker, JP. 2006. The social movement of meat in Taimyr, Northern Russia. Nomadic Peoples 10 (2): 105–122.

Citations

CHAPTER 1

Dana L. K. Hoag and Jay Parsons, "Risk Navigator SRM: An Applied Risk Management Tool,"Journal of Probability and Statistics, vol. 2010, Article ID 214358, 17 pages, 2010. doi:10.1155/2010/214358.

CHAPTER 2

Brian A Child, Jessica Musengezi, Gregory D Parent, and Graham F T Child, The economics and institutional economics of wildlife on private land in Africa, doi:10.1186/2041-7136-2-18.

CHAPTER 3

Dhan Prakash, Sonika Verma, Ranjana Bhatia, and B. N. Tiwary, "Risks and Precautions of Genetically Modified Organisms," ISRN Ecology, vol. 2011, Article ID 369573, 13 pages, 2011. doi:10.5402/2011/369573.

CHAPTER 4

Nick Pidgeon and Baruch Fischhoff, The Role of Social and Decision Sciences in Communicating Uncertain Climate Risks, doi: 10.1038/nclimate1080.

CHAPTER 5

Pablo Rubén Mariconda, Technological risks, transgenic agriculture and alternatives, http://dx.doi.org/10.1590/S1678-31662014000400005

CHAPTER 6

Francis EO Opiyo, Oliver V Wasonga, and Moses M Nyangito, Measuring Household Vulnerability to Climate-Induced Stresses in Pastoral Rangelands of Kenya: Implications for Resilience Programming, doi:10.1186/s13570-014-0010-9.

CHAPTER 7

Eranga K Galappaththi and Fikret Berkes, Institutions for Managing Common-Pool Resources: the Case of Community-Based Shrimp Aquaculture in Northwestern Sri Lanka, doi: 10.1186/s40152-014-0013-6.

CHAPTER 8

Gabriele Volpato and Patricia Howard, The material and cultural recovery of camels and camel husbandry among Sahrawi refugees of Western Sahara, doi: 10.1186/s13570-014-0007-4.

Index

M

N

O

P

Q

R

S

T

U

W